Praise for *This Way to the Universe*

"This book is a rare event: a grand overview of the leading ideas in modern fundamental physics, presented by someone who is a true master. Michael Dine has a well-deserved reputation for being a leading theorist who deeply cares about, and understands, the details of experiments and observations. This will be a rare combination of profound insight and empirical grounding that will delight a broad audience."

> —Sean Carroll, theoretical physicist at the California Institute
> of Technology; host of the *Mindscape* podcast; and author of
> *From Eternity to Here, The Particle at the End of the Universe,*
> and *The Big Picture*

"This book, written by one of the great masters of modern physics, is an extraordinary journey into what we know, what we hope to know, and what we don't know, about the universe and the laws that govern it. Unlike other books on the subject, it does not try to sell you on the author's pet theory. In a clear and honest way, it lays out all the most important problems, puzzles, and potential solutions that modern physics faces."

> —Leonard Susskind, professor of theoretical physics at
> Stanford University and author of *The Cosmic Landscape,*
> *The Black Hole War,* and The Theoretical Minimum book series

THIS WAY
TO THE
UNIVERSE

THIS WAY
TO THE
UNIVERSE

a theoretical physicist's journey

to the edge of reality

MICHAEL DINE

DUTTON

DUTTON

An imprint of Penguin Random House LLC
penguinrandomhouse.com

LIBRARY OF CONGRESS CATALOGING-IN-PUBLICATION DATA
has been applied for.

ISBN 9780593184646 (hardcover)
ISBN 9780593184660 (ebook)

Printed in the United States of America
1 3 5 7 9 10 8 6 4 2

This book is dedicated to
Melanie, Aviva, Jeremy, Shifrah, Matt, and Oren

CONTENTS

CONTENTS

AND STEPPING INTO THE UNSTABLE

THIS WAY
TO THE
UNIVERSE

STEP ONE

CHAPTER 1

SURVEYING THE UNIVERSE

It seems to be an extraordinary moment. On the one hand, we face daunting challenges: climate change, global pandemics, the threat of nuclear war. On the other hand, as a species, we have knowledge of the world—and the universe—around us beyond anything humans might have imagined even a century ago. No matter what happens, we have an unprecedented understanding of the natural world of which our daily experiences sample only a tiny corner. Our lives play out on scales of centimeters, meters, kilometers, perhaps thousands of kilometers. But we know about nature on smaller scales—far, far smaller than the size of an atomic nucleus. We also know about the universe out to unimaginably large distances. Even more amazing is what we know—really know—about events billions of years ago, and we can make statements with near certainty about what will happen to the universe for the next few tens of billions of years. An extraordinary moment indeed.

Most of us have heard about faraway stars and galaxies,

have some inkling that the universe emerged from a big bang billions of years ago. But precisely how large and how old is the universe? Where did it come from? What is its ultimate fate? How do we find answers to these questions?

We are aware of atoms and maybe somewhat aware of things smaller than atoms. How can we possibly know about atomic nuclei that are far too small to see with the most powerful microscopes? How do these tiny things control the operation of the universe at large, as well as events like making a sandwich, using a credit card, or driving to work? From the largest scales to the smallest scales, our universe can seem impossibly mysterious. Can we do more than speculate about the architecture of the cosmos and its building materials? Can we construct experiments that will answer our questions about reality at such fantastic scales?

As I write this, we are still confronting the Covid-19 pandemic. From this ordeal, we're all now familiar with the significance of powers of 10. In the early stages of the outbreak, the number of cases was growing by nearly a factor of 10 every week. Here is what that meant for projections about cases in the United States.

March 2, 2020	100
March 10, 2020	1,000
March 18, 2020	10,000
March 25, 2020	100,000
April 3, 2020	1,000,000
April 7, 2020	10,000,000

From 100 people to 10 million people sick in a matter of five weeks. After that, in this case, the growth would have slowed, only because it would have been harder for the virus to encounter people who had not already been infected. Fortunately, states and local communities, to a large extent, adopted shelter-in-place restrictions within a few days of March 11, 2020, a little less than two weeks after the exponential growth began. Two weeks later, about the time from exposure to the virus to visible symptoms, the effects of the partial lockdown started to be felt. So on March 10, there were 994 cases, just under our thousand expected, 9,307 on March 18, somewhat further below our 10,000 expected. But by March 25, the effects of social distancing became visible, with 68,905 cases. On April 3, 250,000 cases, and on April 11, 509,000—a factor of 200 less than our worst-case scenario. The drastic measures we took as a society saved millions of lives. Had we acted earlier, even more would have been saved; had we waited longer, an even greater catastrophe would have unfolded. Indeed, states and localities that acted earlier generally did better. Around the world, similar stories played out. Subsequent months saw waxing and waning of the virus, tied to behaviors, improved treatment strategies, and the eventual rollout of vaccines.

But powers of 10 need not always tell such a grim story. They are a valuable tool for thinking about nature. We humans occupy a tiny planet in a vast universe. At the same time, there is a world of far tinier things—molecules, atoms, protons, neutrons, and electrons. Powers of 10 are also a useful concept in these happier pursuits. In 1977, while a graduate student visiting the Smithsonian Institution with my brother, I watched the video *Powers of Ten,* by Charles and Ray Eames (a couple best known for their work in industrial design). This beautiful film summarized our understanding of nature at that time, on the largest and smallest scales. Starting with a couple enjoying a beautiful spring day, occupying a space maybe two meters across in each direction, it explored scales progressively larger by factors of 10—parks, cities, states, nations, the planet, the solar system, galaxies, and clusters of galaxies. It then proceeded in the other direction, describing smaller scales—parts of the human anatomy, then cells, atoms, and the nuclei of atoms. It summarized for me pretty well what I was learning in my studies. To be honest, there was a good deal I didn't yet know in that film.

A lot has happened in the subsequent decades. We understand nature at scales several powers of 10 larger and several powers of 10 smaller, and we have clues to many more powers of 10 in each direction. I have been a witness to, and in some cases a participant in, many of these developments. Telling the story of nature on this vast range of scales is the subject of this book. The story follows this progress in physics, astrophysics, and cosmology. I will only occasionally mention the spectacular discoveries of the past century in biology, medicine, computer and cognitive science, and other fields.

These advances have been the product of dedicated work of experimenters and theorists. The dichotomy between the two can be a confusing one, but one which will, I hope, become clear in these pages. While I seriously considered a career in experimental physics, as a student I fell in love with theoretical physics. This was, professionally, a risky choice, and some of my mentors discouraged me, telling me that the competition was just too stiff. While I believed them, and was by no means convinced I had the stuff for theory, I was in love with the subject. My graduate student days were spent studying phenomena at the smallest scales then accessible, about one-third the size of an atomic nucleus, or 10^{-14} centimeters (a hundredth of a trillionth of a centimeter). I have to confess that I was hardly a brilliant student, but my teachers had faith in me, and I went on to do a postdoctoral fellowship at the Stanford Linear Accelerator Center in Menlo Park, California. Here, I was involved in interpreting experiments on still smaller length scales. Among my mentors were Sidney Drell, a leading voice for nuclear arms control, and Leonard Susskind, then a brash young theorist, who had recently come to Stanford. Still, though, I had a hard time finding my way in that period. The problems I worked on didn't really move me.

After two years at Stanford, I moved to a similar position at the Institute for Advanced Study in Princeton. The Institute *is* an institution exclusively devoted to theory, and it is famed, in part, because of the faculty of its early days. This includes, most notably, Albert Einstein, but also figures like J. Robert Oppenheimer (who headed the atomic bomb effort at Los Alamos during the Second World War), John von Neumann (an early computer pioneer), and George Kennan (the diplomat who shaped much of US policy relative to the Soviet Union in the early days of the Cold War). Its current faculty—including Edward Witten, Nathan Seiberg, Juan Maldacena, and Nima Arkani-Hamed—are among the premier living theoretical physicists in the world. In this environment, I found my scientific bearings and started to probe questions beyond the level of our then current understanding. From there, I went on to five productive years on the faculty of the City College of New York, before moving, for family reasons, back to the West Coast, joining the faculty at the University of California at Santa Cruz. I have spent the subsequent three decades there.

UC Santa Cruz sits amidst towering redwoods, overlooking Monterey Bay. When first established in 1965, it was committed to a radical, 1960s vision of education and engagement. Its unofficial motto was "We are not Berkeley," meaning that its faculty and administration were committed to their students, not just to research. That vision survives, but for serendipitous reasons, UCSC also became a research powerhouse. The astronomy enterprise of the UC system, the Lick Observatory, moved its headquarters to the Santa Cruz campus from Mount Hamilton. Earth scientists were attracted to the campus by its proximity to major fault systems, marine biologists by the rich ecology of the nearby bay; biologists, chemists, and mathematicians were excited by the opportunity to work in the natural beauty of this landscape. UCSC also became a center for particle physics, because of the commissioning of a revolutionary new instrument for particle physics nearby at the Stanford Linear Accelerator Center.

I arrived much later, in 1990, imagining a hippy-dippy institution in the woods. And so it was, but I discovered at the same time a rich intellectual and scientific environment. For the same personal reasons that brought me to Santa Cruz, I actually lived "over the hill" on the other side of the Santa Cruz Mountains, in San Jose, part of the Silicon Valley. Fortunately, from the very start, I had a car pool with a group of colleagues. At the time, this group included four high-energy physicists, working on experiments at the Stanford Linear Accelerator Center, the Fermi National Accelerator Laboratory (Fermilab, located near Chicago), and CERN, the big European laboratory in Geneva. There were also two astronomers. Two of the experimenters were playing leading roles in the Superconducting Super Collider (SSC), planned as the world's largest particle accelerator, then in the initial stages of construction near Dallas, Texas. It was designed to accelerate two beams of protons to enormous energies, and then to smash them together, examining the products of the collisions. In a multibillion-dollar project involving thousands of PhD scientists, my car pool partners had the principal responsibility for tracking particles just as they emerged from the collision. One of the astronomers was working on understanding planets. At that time, the existence of planets beyond our solar system was a matter of speculation. All that changed starting in 1995 with the first discovery of an extrasolar planet. Astronomers at Santa Cruz made crucial contributions to the breakthrough technology and to the underlying planetary theory. The other astronomer was a cosmologist, one of the originators of the theory of how dark matter led to the formation of the stars and galaxies.

In 1993, President Bill Clinton recognized that as the costs of the SSC rose, it became more and more vulnerable to the surging politics of government spending. Congress finally killed the project one day in the fall. I expected my colleagues to spend a few days mourning, but by the next morning, they were in the car discussing a call they had received from the big laboratory in Geneva, Switzerland, inviting them to join a project there, known as the Large Hadron Collider (LHC), still in early stages of development. They agreed, and immediately launched into work on the development of a large detector for elementary particles, known as ATLAS. It would be fifteen years before this machine was ready to operate. There were many successes and setbacks on the way, involving science, technology, and funding. The most devastating was a magnet failure, in 2008, which greatly damaged the machine. The recovery took two years, but by 2010, the accelerator was up and working well. In 2012, two experimental teams at the LHC discovered the Higgs particle.

As a theorist, my work involves, among other things, trying to understand the results of experiments and to anticipate possibilities for future experiments. My close connections with experimental colleagues have helped keep me honest, focused on questions we can really hope to answer in experimentally verifiable ways—or at least to distinguish those we can and those we can't. Much of my research effort is devoted to sorting out precisely these issues. What might account for the mass of the Higgs boson? What might the dark matter consist of and under what circumstances might we hope to find it? Is string theory subject to experimental test? Our conversations in the car were often about our children, restaurants, sports, and politics (real and academic), but they were mostly about science. Just as my car pool partners have tried to teach me about the challenges of building electronics that can withstand intense bursts of radiation, they have suffered through my explanations of the latest theoretical ideas and their promise and limitations.

My students at UCSC have also kept me focused on what's exciting in science. I frequently teach a course with the title "Modern Physics." It starts with Einstein and relativity, moves to the development of quantum mechanics, and then proceeds through the spectacular developments of the twentieth and early twenty-first centuries. This book will cover an even broader sweep of ideas, and my hope is to convey excitement about what we understand, and appreciation of the mysteries we currently confront.

The discovery of the Higgs particle, of dark matter and dark energy, along with precision studies of the big bang, illustrate an understanding of our place in the universe beyond anything that humanity has ever known. At the same time, we have burning questions. For some of these, we have a clear path to answers, for others less so. I firmly believe that this science is not so far removed from the ordinary events of our lives that we can't all share in both the understanding and the most pressing questions. I intend to illuminate what questions are likely to be resolved, say, over the next decade, by experiments or new theories, and which ones may not be accessible.

This book will explore many orders of magnitude beyond those which the creators of the *Powers of Ten* video could contemplate. We will journey across scales both voluminous and microminiature, but we'll also travel the scales of time. Our clock will start, t=0, at the big bang. On this clock, our present instant is about 13 billion years later, or 13×10^9 years. From our present moment, we'll look back to times when stars and galaxies began to form—1 billion years after the big bang, and further to the earliest times we completely understand—three minutes after the big bang, when hydrogen and helium were produced in a hot, cosmic soup. But we'll look back much earlier—to times for which we have only scattered bits of evidence, when the universe was perhaps a billionth of a second old, when matter itself may have been created. Ultimately, we'll peek behind the curtain of the big bang, asking what may have come before, and encounter controversial ideas like the *multiverse*. This idea provides a compelling explanation of one—and maybe more than one—of nature's greatest mysteries. It is even conceivable that we could find observational evidence for this bizarre possibility.

Experiment and Theory

Perhaps more than most sciences, physics is riven. This may sound bad, but it has enabled principled, disciplined investigations of much that initially seemed utterly bizarre. Physicists divide into two groups: those who devote the bulk of their time to designing, building, running, and analyzing the data from experiments and those who spend the bulk of *their* time inventing theories, predicting the results of proposed experiments, and comparing experimental results with theories. Some of these theories are designed to account for existing experimental results, some to account for a mysterious, poorly understood feature of nature and its laws. This was not always true. Newton, who did much to shape the modern field of physics, was both an experimentalist and theorist. He performed important measurements, studying, for example, the properties of light (coming to notoriously wrong conclusions). He invented the calculus, one of the most important tools of the modern theorist or experimentalist, and wrote down the basic laws of motion, as well as a theory of gravitation that survived, virtually unscathed, for almost two hundred years. But by the late nineteenth century, theory had emerged as a specialization, practiced in those days by a rather small group. This reflected, at least in part, the growing technological sophistication of experiments and a similar growth in the mathematical demands of theoretical analysis. Even then, the Scotch physicist James Clerk Maxwell, who put the laws of electricity and magnetism into their final form in the 1860s, also did experiments with color and established the Cavendish Lab at Cambridge University in the latter part of his career. An early notable pure theorist was the Dutch physicist Hendrik Lorentz, who, among many other accomplishments, wrote down a precursor to Einstein's relativity theory and developed an early theory of the electron.

The paradigm of the modern theorist, of course, is Albert Einstein. Einstein burst onto the scene in 1905, with three remarkable pieces of work. The two most famous of these are his theory of special relativity and the photoelectric effect, for which he won the Nobel Prize. The average physics student is less aware of his work on the Brownian motion. But this work did much to establish the reality of atoms, gave a reasonably good estimate of the number of atoms in a cubic centimeter of water (Avogadro's number), and had a profound influence not only in physics but in chemistry and biology. All of these were products of some combination of pure thought and analysis of data from existing experiments. It is a mold that all who call themselves theoretical physicists try to emulate. But Einstein spoke wistfully of his own aspiration of doing both experimental and theoretical physics. Of Newton, he wrote: "Nature, to him, was an open book . . . In one person he combined the experimenter, the theorist, the mechanic, and, not least, the artist in exposition . . . He stands before us certain, and alone: his joy in creation and his minute precision are evident in every word and every figure."

A twentieth-century exception to the theorist/experimentalist dichotomy was Enrico Fermi. Born in Italy in 1901, his early theoretical work in quantum mechanics is essential to chemists' understanding of the periodic table and to the physics of neutrinos. But he also did crucial experiments in nuclear physics, for which he was awarded the Nobel Prize in 1938.

He and his wife, Laura Fermi, went to Stockholm to receive the prize, but did not return to Italy. Instead, fearing persecution in fascist Italy as Laura was Jewish, they continued on to the United States, he taking a position at Columbia University. His experiments at Columbia and later at the University of Chicago were pivotal in the development of nuclear weapons and nuclear power. His students included many of the most important theorists and experimentalists of the post–World War II generation, but none exhibited his combination of talents.

My car pool has not only educated me about a range of experimental issues but has helped keep me honest, focused on questions that either are driven by experiment or can be addressed by experiment. A de facto requirement of membership in this pool is the ability to explain to each other what we are doing.

As we make our journey, we will encounter many individual physicists, both historical figures and theorists and experimentalists active now. We will encounter men and women from different continents, but it is hard to avoid the reality that the field has been dominated by *men* from a handful of countries. Some of our actors have harbored offensive views along racial, ethnic, or gender lines. Nevertheless, I do believe the evidence is strong that the questions we will encounter are of interest across the lines that sometimes separate us, and I hope sharing these can help bring us together.

CHAPTER 2

CAN WE TAKE SPACE
AND TIME FOR GRANTED?

The arena for our day-to-day lives—and for exploring the realms of the very large and the very small—is normally described in terms of space and time. "I'm sorry, I'm running late," we text. "LA is three hours behind New York, isn't it?" or we might say, "Mt. Everest is only about five miles high, right?" We all have some working intuition for the reality of space and time. But the laws of nature give a sharper meaning to space and time themselves.

Isaac Newton (1643–1727) was a complex person, and his scientific evolution was equally complex. His father died shortly before his birth, and, as a child, he was for a time abandoned by his mother. He was a difficult, hot-tempered individual who formed few lasting relationships. He had strong religious views and dabbled at various points in his life with alchemy. In his later years, he left Cambridge for London and largely gave up his scientific researches to become Master of the Mint. He secured this final position through friends. What might have been a sinecure became his passion. He pursued new, better standardized coinage, but especially threw his energies into the investigation (often undercover), arrest, and execution of counterfeiters—typically by hanging, drawing, and quartering.

It is from Newton, more than his predecessors or even his most notable contemporaries, that we acquire the notion that the phenomena we witness in nature are governed by *laws*, and that these laws can be expressed in precise, mathematical terms. He framed his questions in language shaped by the worldviews of his age. He was influenced, certainly, by Galileo, but also by his *competition* with some of his contemporaries, most notably the English scientist Robert Hooke. Their rivalry was quite fierce. Hooke felt, possibly unjustly, that Newton stole his law of gravitation from him. Newton rejected any such insinuation. He famously said, of his own accomplishments, in a letter to Hooke: "If I have seen further it is by standing on the shoulders of giants." This story is often told as an illustration of scientific modesty, but as one of my astronomy colleagues explained to me, Hooke was quite slight (he actually described him as a dwarf, but this apparently was not true). While some of my colleagues think pretty highly of themselves, few are quite so cruel—often quite the opposite.

Newton, at various stages in his career, worked in chemistry and on phenomena connected with light. He observed correctly that white light is composed of light of different colors, but he not so correctly theorized that light consisted of particles, or "corpuscles." For the questions that concern us, Newton's investigations understanding the motion of the planets and the moon are central.

Newton and the Character of Physical Law

The title of this section is taken, with apologies, from a set of lectures given in 1964 by Richard Feynman. We speak casually of physical law, or laws of nature, but Feynman asked: What do we mean by this? Our modern scientific paradigm for physical law comes from Newton's laws of motion and his law of gravity. Newton's successful application of his rules and techniques to the motion of objects in the solar system was the first great triumph of this worldview.

We may imagine Newton as he described himself, sitting in a garden and watching the fall of an apple. From a modern perspective, we can divide Newton's laws into two types. The first were his laws of motion, a basic framework, a set of rules applicable to a broad array of physical phenomena. The first law of motion says that a body in motion remains in the same state of motion (it moves with the same speed, or velocity) unless acted on by a force. This may seem intuitively reasonable, but begs the question: What's a force? Newton provided a definition, through his second law: A force gives rise to an *acceleration*, a change in speed, or velocity. This law says that, for an object, say a car, of a given mass, the acceleration gets larger if the force is larger. For a given amount of force, the acceleration is *smaller* if the mass is larger. So, for the same force, my Prius C accelerates much more rapidly than a large truck. Similarly, if I press the gas and double the force of the engine, the particle accelerates to highway speeds twice as fast. These two rules account for the motion of things—day-to-day objects like tennis balls, bullets, and missiles, but also enormous bodies like planets, stars, and galaxies.

Having established this framework for motion in space and time (what is called *kinematics*), Newton took another crucial step, determining the force of gravity. Humans had observed the motion of the planets since ancient times. In the century before Newton, a more precise picture had developed of planetary motion. The sixteenth-century Danish astronomer Tycho Brahe (1546–1601) did careful, accurate measurements of the planetary orbits in an observatory funded by the king of Denmark. His measurements were particularly remarkable as he did not have a telescope, instead relying on instruments of ancient origin, the sextant and the quadrant, to measure the positions of planets and stars in the sky. Johannes Kepler (1571–1630), Brahe's assistant, analyzed the data, summarizing it in three "laws." One characterized the shape of the orbits. A second governed the speeds of the planets as they move through their orbits. A third related the length of their year, the time it takes them to complete an orbit, to their distance from the sun. These rules were remarkable in that they didn't correspond to any obvious intuition and they went against Kepler's prejudices at the start of the project. They are also remarkable in the care they show. Kepler could have declared the orbits to be circles, and, in fact, the shape of the orbits of all the planets then known are *nearly* circles, but not quite; they are actually el-lipses. Before the demotion of Pluto from its status as planet to dwarf planet, I liked to remark to my students that it was the only planet whose orbit was not very nearly circular.

I put *laws* in quotes above because Kepler's pronounce-ments, while providing a way of organizing Tycho Brahe's ob-servations, were not laws in the sense of the laws that Newton would promulgate. The difference is subtle and beautiful. The first element of this distinction has to do with the broad appli-cability of Newton's laws.

The lesson of the story of Newton and the apple, apocryphal or not, is not about how Newton discovered objects fall down. It is his realization that the force of gravity is *universal*, that the moon, in particular, falls toward the earth according to the same rule that the apple falls toward the earth, as does the earth to-ward the sun. He translated this into the statement that the force, say between the sun and a planet, is proportional to the mass of the sun, times the mass of the planet, divided by the *square* of the distance between them. This means, for example, since Venus has a mass similar to that of the earth but is closer to the sun—about 62 million miles as compared to 98 million—the pull of the sun is roughly twice as large on Venus as on the earth. Correspondingly, Venus moves much faster than earth in its orbit, while the outer planets move more slowly. The force of gravity the earth exerts on the apple is smaller than it exerts on the moon since the apple has far less mass, and it would not be appreciable were it not for the much shorter distance from the center of the earth to the apple compared to the moon. With this law for the force, Newton could explain, indeed, all three of Kepler's laws. But Newton's laws were much more powerful than Kepler's, making much more precise predictions. Newton realized that the orbits are *not exactly* ellipses. The planets are pulled on not only by the sun but by the other planets and their moons as well. Because the planets are much smaller than the sun, these effects are small, but Newton's work opened up the possibility of a far more detailed study of planetary motion, which continues to this day. It was not until two centuries later that Einstein, with his theory of general relativity, predicted tiny breakings of the law, which were confirmed by experiment.

Returning to the distinction between Kepler and Newton, Kepler's laws summarized regularities he observed in data. They were not exact (though he likely believed so), and, more importantly, Kepler could not say, a priori, just how precise his rules should be. Newton *could* explain the small violations of Kepler's rules. Einstein's relativity principles enlarged and transcended both Newton's laws of motion and his law of gravity, and would allow him to say when *Newton's* laws should be valid and when they would not apply.

These issues aside, Newton's kinematic framework became the bedrock of a range of technologies. Even when the underlying forces could not be described in as simple a way as the force of gravity between the sun and the planets, problems ranging from civil engineering challenges to motion of projectiles (artillery shells, missiles), features of weather, and much, much more could—and still can—be described in terms of Newton's laws of motion.

Newton, Space, and Time

Most of us live surrounded by devices that keep track of time and where we are in space with precision. Forgotten are the days when our clocks and watches were reliable only to within a minute or two. Our phones keep the time accurate to tiny fractions of a second. Our navigation apps predict the time it will take us to cover distances in our cars, on foot or bicycle, reliably, even accounting for the volume of traffic and our individual pace or driving style. Newton was crucial to a transition from standards of time and space tied to our daily experience to those tied to nature's laws.

Historically, measurements of distance seem to have first been tied to human anatomy—the foot, with an obvious origin, and the cubit, which was the length of the forearm. The mile was originally the distance covered by a Roman soldier in 1,000 steps. These clearly suffered from issues of standardization. Miles would have different lengths depending on how tall, well fed, and rested the soldiers were. Much later, the meter would be defined, first, as one 10-millionth of the distance from the north pole to the equator. This measure, while quite arbitrary, at least has the virtue that everyone can agree, with some small error, as to what this distance is.

Providing a standard for measuring time poses other challenges. The day is an obvious starting point to keep track of the passage of time, and we owe to the ancient Babylonians its division into hours, minutes, and seconds. But the length of the day varies by quite a bit through the year. Eventually, the notion of an average day allowed some standardization of these measures. The lunar month is convenient, but there is not a fixed number of lunar months in a solar year. The year, measured as the time it takes the earth to complete one orbit around the sun, is pretty steady, varying by fractions of a second from year to year (and sporadically adjusted by the insertion of a "leap second"). Of course, it was only relatively late in human history that reliable measurements of the earth's position in its orbit became commonplace.

Galileo took the measurement of time to new levels, studying the motion of pendula, and discovering that the time it takes a pendulum to make a single swing, if you give it a small push, just depends on its length and not how large the swing is. With this "law," one could reliably measure the passage of time by counting the number of pendulum swings. By comparing the lengths of their pendula, two timekeepers could agree how much time had passed.

This is a big step toward the concept of time that Newton gave us. Time and space were the *setting* for his laws of motion and his laws, but the operation of the laws provided a definition of time itself. This was a dramatic generalization of Galileo's observations of pendula.

It is perhaps not surprising that we associate the notion of *law of nature* with Newton. The concept of law fit with his personality, which could be quite rigid. He would assert that things were as he said they were, and not tolerate criticism or disagreement. This lack of tolerance for divergent views is apparent in his description of time. Time, for Newton, was absolute, a concept whose meaning he took to be self-evident and not subject to question. One might argue that he tried to shut down any debate on the subject of time or space and their meaning when he wrote:

> Absolute, true and mathematical time, of itself and from its own nature, flows equably without relation to anything external. Absolute space, in its own nature, without relation to anything stable, remains always similar and immovable.

Just in case, Newton invoked God as his authority:

> Absolute time is not an object of perception. The Deity endures forever and is everywhere present, and by existing always and everywhere, He constitutes duration and space.

Newton is, perhaps, being modest here. If we substitute "Newton's laws" for the "deity," these statements have some truth. Newton's laws of motion are bound to the nature of time and space, and simultaneously give them a definition. Newton could readily use his laws to understand Galileo's "law" for pendula and generalize this to other types of clocks. If one knows the force, say on the spring in an old-fashioned clock, or the behavior of atoms inside your cell phone's clock, counting the oscillations back and forth of the spring or of the atoms, one has another measure of time. Rather than time or space as absolute, to the extent that Newton's laws are *absolutely true and exact*, they provide a definition of time.

Leaving God aside, what proves that the passage of time is the same everywhere and always? Indeed, this question was raised by Ernst Mach, a physicist and philosopher who was active in Austria in the latter part of the nineteenth century. One of Einstein's intellectual heroes, Mach wrote of Newton's insistence on absolute time: "Absolute time is a useless metaphysical concept, and cannot be produced in experience. Newton acted contrary to his expressed intention only to investigate actual facts." But if Newton's laws define this absolute time, unless something is wrong with the laws, absolute time is just fine. For the better part of two centuries, Newton's laws of motion and gravity remained safely enthroned. The first challenges were hidden in the law of electromagnetism, codified by James Clerk Maxwell, and uncovered by Albert Einstein.

More Laws of Nature: James Clerk Maxwell

Gravity features heavily in our day-to-day lives, from anchoring us to the earth, to holding the earth in its orbit around the sun and the moon in its orbit around the earth. So it is perhaps not a surprise that gravitation was the first realm in nature to be understood as governed by physical law. But there is another class of phenomena that are at least as important: those of electricity and magnetism. After Newton, it was natural for scientists to seek a setting similar to the law of gravitation for these phenomena. But developing a complete picture took a good part of two centuries.

One obstacle to progress lies in my claim that electricity and magnetism dominate our day-to-day lives. This is true, but hardly obvious to most of us today, and certainly not so even at the beginning of the twentieth century. Electrical forces hold electrons to atomic nuclei in atoms, and are responsible for the structure of matter and for all of chemistry. Electrical signals control the motion of our bodies, and the force of friction (another controlling element in our daily existence) arises from the small electrical forces between neutral atoms. Magnets, known to most of us from early childhood, are the result of very complex phenomena involving electrons in special materials, such as iron. One can't really get started understanding them without quantum mechanics, a subject we will turn to later, which didn't really open up until the 1920s. And what is, in some ways, least obvious is that light, radio waves, the microwaves we cook with, and the X-rays used by our doctors and dentists are all the result of electricity and magnetism working together in concert—what is called *electromagnetism.*

Only slowly did the pieces of this puzzle come together. Benjamin Franklin, among others, did experiments establishing that electricity is the result of the motion of electric charges. The French engineer/scientist Charles-Augustin de Coulomb established in the late 1700s that charged objects attract or repel each other according to a law similar in many ways to Newton's law of gravity. Indeed, the main differences between Coulomb's law and Newton's are in the strength of the force and that, in Newton's theory, all massive objects attract each other, whereas in Coulomb's, objects with the opposite sign of the charge attract each other, but those of the same sign, repel. It is this distinction between gravitational and electric forces that inspires the science fiction of antigravity.

The connection between electricity and magnetism was first discovered by the English scientist Michael Faraday in the early 1800s, not so long after Coulomb. Faraday formulated the notion of electric and magnetic *fields*. The idea was that an electric charge is surrounded by an *electric* field. A charged particle that passes through such a field experiences a force, which grows or shrinks as the field gets larger or smaller. A charged particle moving through space produces a *magnetic* field. As a result, a wire carrying electric current produces such a field. A charged particle like an electron passing through a magnetic field also experiences a force, but the force is more complicated (the bane of existence of many an undergraduate student), depending not only on the strength of the field but also on the speed at which the particle is moving.

For Faraday, the fields were a useful device to picture the effects of electric charges, but they were simply the slaves of the charges and currents; they had no independent existence. This changed in 1865 with the work of James Clerk Maxwell, a Scottish physicist. In Einstein's words,

> A new concept appeared in physics, the most important invention since Newton's time: the field. It needed great scientific imagination to realize that it is not the charges nor the particles but the field in the space between the charges and the particles that is essential for the description of physical phenomena.

The field concept proved successful when it led to the formulation of Maxwell's equations describing the structure of the electromagnetic field.

Maxwell took the equations of Coulomb and Faraday (and also André-Marie Ampère, for whom amps are named) but realized they did not make complete sense. He added another term, and with this discovered that electricity and magnetism are, indeed, responsible for electromagnetic radiation. This *explained* light—light comes in waves of electric and magnetic fields, as opposed to Newton's "corpuscles." Dramatic in itself, Maxwell's theory predicted that such radiation could come in many different forms. Specifically, he predicted the existence of radio waves, which were subsequently discovered by Heinrich Hertz (for whom the unit of frequency we hear on the radio is named). There was, finally, a unified picture of electricity and magnetism, and the field concept was a crucial component.

Einstein and the Fall of Absolute Time

As Mach said, Newton's insistence on the absolute nature of space and time lay on a shaky foundation. But Newton's laws of motion and gravity at least held together, whatever one's philosophical objections. Maxwell's equations, however, contained the undoing of Newton's absolute time, and of space as well.

The puzzle created by Maxwell's theory was connected to the speed of light. One triumph of the theory was that this speed could be related to other, measured quantities. But the problem was that according to the equations, light always moves at the same speed, *no matter what.* To physicists of the late nineteenth century, this made no sense. They reasoned that light waves were like water waves. If you throw a stone in the water, it creates waves that move away from you at a certain speed. If you throw a stone from a moving boat, the waves move away from you roughly at the same speed as if the boat were still. So as far as someone watching from shore is concerned, they see the waves moving even faster. But Maxwell's equations didn't seem to allow for this. Light waves produced by a lamp in your hand move away from you as fast as light waves from a flashlight on a fast rocket ship move away from you.

Maxwell and his contemporaries were not really troubled by this. They couldn't imagine that light waves were entities unto themselves. Instead, they believed that, like water waves, which are disturbances propagating through the liquid, light was some sort of disturbance of a medium, which they called the *aether.* This aether permeated all of space. The velocity of light of Maxwell's equations, they believed, was the velocity relative to this aether. But the aether hypothesis was already in trouble by 1887 (before Einstein put forward his theory). Albert A. Michelson and Edward W. Morley of Case Western Reserve University in Cleveland reasoned that one should be able to see evidence that the earth moves relative to the aether. In a famous experiment, they could find no support for the aether hypothesis. To what extent this experiment influenced Einstein is unclear, and perhaps not so important.

When Einstein seemingly spoke of Maxwell's "great scientific imagination," he was perhaps more accurately describing himself. Einstein shed the artifice of the aether and accepted the fields as entities in their own right. He asserted that the speed of light is a quantity intrinsic to the fields. This resolved the dilemma of Maxwell's equations and the aether, but at the cost of absolute space and time. The possibility was anticipated by the great French mathematician Henri Poincaré, who, struggling to reconcile the Michelson-Morley result with the notions of an aether, wrote, "Not only do we have no direct intuition of the equality of two times, we do not even have one of the simultaneity of two events occurring in different places." Poincaré *almost* had Einstein's relativity, but he was too wedded to the aether concept to take the final step.

Einstein's Annus Mirabilis

In 1905, Einstein, then twenty-six years old and underemployed in the Swiss patent office in Bern, took his three huge steps. First, the development of his theory of special relativity. If you have not taken several years of physics courses, you can be excused if you feel uncomfortable when the subject of Einstein's relativity theory is mentioned. The first source of confusion is that there are *two* theories known as Einstein's relativity theory, the *special theory of relativity* and the *general theory of relativity*. The logic of the names is a bit obscure; they are quite different types of theories, and the general theory is not simply a generalization of the special theory. The special theory has been well understood and experimentally tested for more than a century. The general theory is much more challenging to confirm, though by now the evidence in support of the theory is compelling. It will be the subject of the next chapter. It is the special theory that will concern us here.

The special theory of relativity forces a dramatic modification of Newton's conception of absolute time and space. Einstein took Maxwell's equations at their word and asserted that the speed of light is absolute: Any observer, whether moving or stationary relative to its source, measures the same speed, usually denoted by the letter c, and equal to 186,000 miles per second (300,000 kilometers per second). But now there's a problem. Suppose there are three passengers in a moving railway car, one in the middle, one in the front, and one in the back. The passenger in the middle flashes a lamp on and off. The passengers at the two ends, looking at their watches, receive a ray of light from the lamp at the same time. The problem is that from the perspective of an observer on the ground, because the passengers are moving, the light has to travel farther to catch up with the passenger in front than the one in the rear. So, since she sees the light rays as moving with the same speed, she sees it reach the passenger in front first. This is *exactly* Poincaré's challenge. An event that is simultaneous for passengers on the train is not simultaneous for an observer on the ground. The very notion of simultaneity is *relative.* But not everything is relative; everyone agrees on the speed of light.

Einstein made all of this precise, with equations that related the intervals of time and space measured by observers moving relative to one another. Particularly striking is that Einstein's principle of relativity mixes up space and time. What we mean by time depends on how we're moving and where we are. It requires that space and time be viewed as a unified entity— space-time. Rather than saying we live in three dimensions, we are forced to acknowledge that we live in *four*, where time is the fourth dimension. Time is no longer absolute; it passes differently for different observers. Time runs more slowly for me, for example, if I'm sitting in a very fast rocket ship, than for an observer sitting still on the ground watching me race by. Just as unsettling, the very idea that two events are simultaneous is relative.

Notions like energy and momentum are also relative. In order that the principle be satisfied, Einstein's special relativity said that a particle at rest has energy, $E=mc^2$, one of the most famous equations in all of science. He worked out, more generally, a new version of Newton's kinematics. These rules have been tested experimentally with extraordinary precision.

On the other hand, it should be stressed that Newton's laws are correct and accurate in situations where the motion of objects is small compared to the speed of light, c. Traveling at the speed of light, one could circle the earth about 8 times per second, or reach the moon in about a second and a half. None of us experience anything like these speeds. Even the fastest rocket needs a bit over an hour to cover the distance light travels in a second.

Perhaps more interesting is the speed of an electron in an atom, which is about 1 percent of the speed of light. The effects of relativity in atoms are very small but are very well measured and are in agreement with Einstein's theory. In modern particle accelerators particles moved at nearly the speed of light, and the same is true for objects we observe in the cosmos; in both cases, relativity explains so much of what we see.

Special relativity represented the first dramatic modification of Newton's picture of space and time. The second came with Einstein's general relativity a decade later, in which space and time are altered by the presence of large amounts of mass or energy—a star, galaxy, or black hole. Einstein's special relativity, while a major alternation of Newton's kinematics, left the basic Newtonian framework of physics intact. This is true of general relativity as well, but the modifications are even more dramatic, as we'll see. The structure that Newton and Einstein bequeathed to us is today referred to as *classical physics*. A much more drastic upheaval was in store, with quantum mechanics, which Einstein also helped to set in motion in this miraculous year. But this revolution would take some time. Let's first dig into the genius of general relativity.

CHAPTER 3

WHAT DO WE MEAN
BY *UNIVERSE?*

In 1905, physicists understood something of the laws governing two types of forces: those of electricity and magnetism
and of gravity. We've seen that the laws of electricity and magnetism, encoded in Maxwell's equations, forced a rethinking of
the most basic concepts of space and time. But what about
Newton's law of gravity? How might it be purged of the notions
of absolute space and absolute time?

In 1907, two years after putting forward his special relativity, Einstein was asked to write a review of the subject. In the
course of this project, he confronted the question: How does
Newton's theory of gravity fit in with his principles? The simple answer: It doesn't. This was actually related to a deficiency
of Newton's law of gravitation that was clear from the moment
he promulgated it. Newton—and perhaps more importantly
his critics—were very troubled by a feature of his theory called
action at a distance. In Newton's laws, if, say, the sun were

suddenly to "jump" (for the moment, take the far-fetched possibility of some space invader attaching rockets to it), the effect on the planets in the solar system would be instantaneous. This despite the fact that the planets are far away. Neptune, for example, is so far away that it takes light from the sun four hours to reach it, but it would move immediately in response to the sun's sudden motion. Newton was criticized for this—was he suggesting that some higher being was responsible for the force between stars and planets? But his law was extremely successful, and, for almost two centuries, this question was largely put aside. Indeed, it was only in the early twentieth century that it was possible to seriously test this disturbing feature.

But with special relativity, one could no longer look away. It didn't make sense that this principle should apply to electromagnetism but not to the force of gravity. It was hard to see how Einstein's assertion, that events at one place and time can affect events at another only after at least enough time has passed for light to travel from one to the other, would not have to apply to any law of nature. This was not a crisis in the sense of an obvious experimental or observational problem. Newton's theory works so well because the speed of light is so large. Light moves so fast that in most of the situations that astronomers encountered in the two centuries after Newton put forward his law, the effects of the finite (as opposed to infinite!) speed of the propagation of information and interactions were impossible to see. Still, Einstein began to consider how one might modify Newton's theory so as to retain its enormous successes while accommodating the relativity principle. In other words, the new laws, in situations where the objects under study are moving much more slowly than the speed of light, or in which the force of gravity is not *too* strong, should reduce to Newton's.

The process of arriving at what Einstein called his general theory of relativity involved a struggle of eight years and a combination of extraordinary scientific insight and sheer hard work. Along the way, there were many missteps. But the theory even more fully revealed Einstein's genius than did his accomplishments of 1905. Einstein might have gone at the problem by observing that Newton's gravitational force law is almost the same as Coulomb's force law for charged particles. Just replace electric charges by masses, and they look alike. The electric force is described by Maxwell's equations. So he might have tried to write equations like Maxwell's, but for the gravitational force.

This is how I would likely have proceeded, and the result would have been failure. But Einstein thought much more deeply before starting his struggle. He was struck by the fact that planets, stars, and other celestial objects all pull on each other; they never push each other apart. This is different from electric forces, where while a proton attracts an electron—pulls the electron toward itself—two protons repel each other. The force of gravity always seems to be attractive, never repulsive. This is hard to mimic with Coulomb's law. Einstein instead took his cue from observations that predated Newton.

Among Galileo's most famous experiments were his studies of falling objects. Archimedes, the ancient Greek philosopher, had asserted that heavy objects fall faster than lighter ones.

This was a plausible guess, but not a statement based on careful observations. Galileo was skeptical and studied the question experimentally. Whether he actually dropped objects of different mass from the Leaning Tower of Pisa is a subject of scholarly debate, but he did perform experiments in which he established that objects of different mass fall to earth at the same rate, neglecting the fact that the air tends to slow everything down. (On the surface of the earth, a piece of paper falls much more slowly than a brick, due to the resistance of the air, but you can easily do a version of this experiment dropping two heavy objects, of different weight, from the same height.) These observations had been improved over the intervening centuries by various investigators, including Newton. Very sensitive experiments were conducted in the late nineteenth century by Baron Loránd Eötvös, a Hungarian physicist, who used a different strategy, attaching various objects to a rod. The device was set up so that it would move if objects of different types responded differently to gravity, but not otherwise. Eötvös established that, for a range of substances, these responses were the same to better than a part in a million; present experiments do thousands of times better.

In Newton's laws, mass has to do with inertia, the rate at which things accelerate in response to a force. But it also has to do with the strength of the force of gravity between two objects. Newton, presumably under the influence of Galileo, assumed that these two kinds of mass were the same. But as far as Newton was concerned, this was simply a fact; no deep principle forced this relationship. Eötvös (and others) established that the *inertial mass* is the same as the *gravitational mass* to a high degree of accuracy. Einstein started with this observation and assumed that the equivalence is *exact*. He then performed a very simple but very ingenious thought experiment, in a setting from daily life. In developing special relativity, Einstein had reasoned by analogy with experiences of one important technology of his day—railroads. He now reasoned by invoking another, newer technology—elevators. He imagined cutting an elevator cable so that the elevator would fall freely (a rather scary prospect). He noted that due to this assumed equivalence of inertial and gravitational mass, observers in an elevator in free fall would experience what we now call weightlessness. They could, for example, float around in the elevator, or pass a ball back and forth with no sense of gravity. It would be as if no gravitational force acted on the objects in the elevators. For the passengers, unfortunately, this would last only until the elevator hit the bottom of the shaft. But nowadays we routinely achieve weightlessness in space travel. The International Space Station, when it orbits the earth, is in *free fall*. It falls due to the earth's gravity. It stays in orbit because the downward pull of gravity competes with the energy of motion provided by the initial launch, to keep the spacecraft constantly circling around the earth. The effects of free fall can also be achieved with aircraft by shutting off the engines for a period. This is routinely done as part of astronaut training. Famously, Stephen Hawking, one of the great gravity theorists, was treated to such a flight in 2007.

Einstein didn't have the advantage of this experience, and the tallest buildings of his day would have allowed a fall of only 4 or 5 seconds. But he realized the phenomenon of weightlessness would follow from the observations of Galileo and Eötvös. Einstein called his realization "the happiest thought of my life" and elevated this to a principle: No experiment can distinguish free fall in a gravitational field from motion with uniform acceleration (as in the elevator). He put forward the hypothesis, his "Principle of Equivalence," that this should apply to all laws of nature: gravity, electromagnetism, and laws yet to be discovered.

From here to mathematical equations was a long struggle. Einstein knew roughly what he was looking for, but when he set out on his journey he did not possess a suitable mathematics for achieving it. David Hilbert, a professor in Gottingen, Germany, and one of the greatest mathematicians of the day, did know the required mathematics and was also in a quest for a theory of gravity; it is likely that had he fully understood the physics issues, he would have gotten to general relativity first, and indeed he almost did. In 1915, however, Einstein completed and published his general theory. The theory met his basic requirements. First, it was consistent with the principles of special relativity. For example, the gravitational interaction propagated at the speed of light; there was no action at a distance. Second, it incorporated the principle of equivalence. Finally, it reduced to Newton's laws except in very exceptional circumstances. Around typical stars and planets, the corrections would be very tiny.

Einstein's theory presented a radical new conception of space and time. No longer were they fixed eternally, but they responded to the presence of matter. Space might be curved, time might run faster or slower near larger or smaller concentrations of matter. Most physicists and mathematicians familiar with the theory would describe it as beautiful, but while the principles are simple, the mathematics is rather complicated, and calculations can be challenging. Einstein, however, focused not only on the great principles and the beautiful mathematics but on the observational consequences of the theory. Because in most circumstances the corrections to Newton's laws are extremely tiny, he had to look for situations where these effects, though small, would be sufficiently prominent to be detectable. He made three predictions that one could realistically hope to test with the technologies then available.

One of these predictions might be more properly described as a "postdiction," an explanation of an already known puzzle in the motion of the planet Mercury. The sun exerts the dominant force on each of the planets; the planets also pull on each other, but these effects are relatively small. Taking into account, first, only the force due to the sun, Newton had shown that the planets would move in orbits the shape of ellipses, just as Kepler observed. According to Newton, these orbits should retain their shape and orientation for all time, ignoring the pull of the other planets.

Even in Newton's day, astronomers studied the motion of the planets with precision. They carefully calculated the orbits on paper, making corrections for all sorts of small effects, such as the pull of the planets on each other. They compared these calculations with equally careful observations. They concluded that small corrections due to the other planets and other effects would lead to a slow deviation from Newton's results; the ellipse would gradually rotate over time. This is referred to (by those with a better memory than mine of their high school analytic geometry) as the precession of the perihelion. Already in the 1850s, astronomers noted that the precession of Mercury was not *quite* at the speeds predicted by Newton's laws; there was a tiny deviation. They proposed a variety of explanations, such as a small, unseen planet or dust, but none was compelling.

Einstein was aware of the discrepancy in Mercury's motion. He realized that Mercury, being the planet closest to the sun, experiences the strongest force of gravity and was thus a promising testing ground for his theory. Einstein set out to calculate the correction to Newton's result. He found it was exactly what was needed to account for the observed precession. I can only begin to imagine how he felt. For a physicist, discovery of a new law of nature is the supreme accomplishment. I have speculated as to several, but the likelihood that any one is true is, typically, not high. Einstein indeed recalled that he was enormously excited—he said he had palpitations—and with the correct result for Mercury's perihelion, he became convinced that his theory was correct.

But inventing theories to explain possible observational discrepancies is still within the realm of more "routine" science. Even better was to come. The second prediction was a real prediction in the sense that he proposed a measurement that had not yet been performed and predicted the outcome. In Newton's theory, one describes the force of gravity as acting on mass. The path of a satellite passing near the sun would be bent by the pull of the sun's gravity. But in special relativity, mass is just a form of energy, and in the general theory, gravity acts on all forms of energy. Light has no mass, but it does carry energy. So the paths of light rays should be altered from simple straight lines as they pass near objects with strong gravity. In 1911, before the theory was fully developed, Einstein tried to calculate the effect. He found that one should be able to see a slight alteration in the position of stars lined up with the sun during a solar eclipse.

Einstein was a genius—and he was also lucky. As I said, the mathematics of general relativity is complicated and was, at that time, also rather unfamiliar. It turned out that in his first calculation of the bending of light by the sun, before he had his theory in its final form, Einstein had made a mistake. He actually obtained the value that Newton would have obtained if the energy of the light was treated as equivalent to mass, through $E=mc^2$. Already in 1912, and again in 1914, expeditions to observe the bending of light during eclipses failed to obtain results, the first time due to rain, the second when it was canceled due to the outbreak of the First World War. In 1915, the year in which he published the final version of the general theory, he got the correct result for the bending of light, finding double the Newton value. The war prevented further measurements until 1919. In that year, two expeditions, one to Príncipe island led by the English astronomer Arthur Eddington, and one to Brazil by Andrew Crommelin of the Greenwich observatory, succeeded in observing the effect. The results were announced in a joint meeting of the Royal Society and the Royal Astronomical Society: Einstein's prediction was confirmed. By this time, Einstein was already well known in the scientific community, and occasional articles about him had appeared in the popular press, but now his name became a household word.

The headline of the *London Times* of November 17, 1919, was typical: "Revolution in Science. New Theory of the Universe. Newtonian Ideas Overthrown."

When I was a student, Einstein's theory of general relativity was a subject of fascination—something any self-described theoretical physicist should know something about. At the same time, actually saying that you might *work* on it would lead to rolling of eyes. There was, in those days, only very limited evidence that the theory was correct—beyond the perihelion and the bending of light, only a phenomenon called the *redshift*—and it seemed that only dreamers imagined there would be new tests. Perhaps even worse, the theory, when combined with quantum mechanics, the subject of the next chapter, did not seem to make sense. Attacking *that* problem put you even more on the fringe. Still, most of the great theorists of the era had taken a stab at these issues, including Richard Feynman and Lev Landau (one of the greatest of twentieth-century Russian theoretical physicists). In the 1980s, perhaps more famously, Stephen Hawking raised issues that challenged the notion that general relativity and quantum mechanics *could* be reconciled and argued that a reformulation of quantum mechanics would be necessary.

Over the course of my career, all that has changed dramatically. Einstein's theory is now a well-tested theory. Our understanding of general relativity is an important tool in our explorations of the universe. Observation of black holes is almost routine. General relativity is a crucial tool in determining the composition of the present universe, and essential, as we'll see, to our understanding of the big bang. Recently, the discovery of the gravitational waves predicted by the theory over a century ago has opened up a new window on astrophysical phenomena. General relativity even plays a role in our navigation apps (through the Global Positioning System, or GPS). On the quantum mechanical side, we have learned a great deal as well, although experimental verification of what we do understand (and clues as to what we don't) is probably not around the corner.

Holding On to Time in General Relativity

The first two tests that Einstein put forward in his 1915 paper are those we've described above: the precession of Mercury's orbit, and the bending of light by the sun. The third had to do quite directly with time. In special relativity, two observers, moving relative to one another, not only can't agree what time it is, they can't agree that two events happened at the same time. In Einstein's general relativity, the situation becomes more extreme. In a gravitational field, near a massive star, for example, time runs more slowly. This effect, known as the *gravitational redshift*, was first observed in experiments by Robert Pound and Glen Rebka in 1959. The effects on earth are very tiny. Pound and Rebka measured the change in the frequency—the number of beats per second—in a particular atomic process due to gravity. The effect was just one part in 10 million! Think of the atom as a clock that ticks 2×10^{19} times per second, or once every 5×10^{-20} seconds (5 100-millionths of a trillionth of a second). Pound and Rebka, in an ingenious experiment performed in a building at Harvard University, observed a change of about 10^{12} ticks in a second. This is a change in the timing of the ticks of about 10^{-26} seconds! At the surface of the sun, the strength of gravity is about 3,000 times what it is on earth, so the slowing of time is about a part in a thousand.

The effects in the gravitational field near a neutron star would be much more substantial. Neutron stars are the debris of supernova explosions, some of the most dramatic events in the cosmos. They typically have a mass about the mass of our sun, packed into a sphere with a radius of a kilometer (just over half a mile; the radius of our sun is about 700,000 kilometers (430,000 miles). So a neutron star is about 10^{18} times as dense as the sun—it's called a neutron star because it is essentially a collection of neutrons packed close together. On the surface of a neutron star, a gram of water (about a teaspoon) would weigh about 10,000 tons (about a billion times as much as on the surface of the earth). In these circumstances, the slowing of time is really dramatic. What normally takes an hour might take two hours. In fact, depending on the precise mass of the neutron star, it might take *much* longer.

This is only the beginning of the drama. A neutron star is on the edge of being a black hole, an object around and within which space and time truly get scrambled. Indeed, black holes with about the mass of the sun probably formed in stellar collapse in the same way that neutron stars do.

Near a neutron star, your survival would really be problematic. Not only would you weigh a billion times more than you do on the surface of the earth, but the force of gravity on your feet would be far greater than on your head. You would be ripped apart by a force in excess of millions of tons.

Don't worry. We're not going to be traveling to neutron stars any time soon, and if one was able to pass nearby, one would surely take precautions to avoid getting too close. But apart from being the stuff of science fiction, this example demonstrates just how extreme the effects of gravity can be. The tiny effects of bending of light near the surface of the sun, in particular, are hugely amplified. A beam of light starting out at the surface of a neutron star would experience such a strong pull of gravity that it would barely make it out.

A black hole is an even more extreme environment than a neutron star. Black holes were first considered by Robert Oppenheimer in 1939, who was then at University of California at Berkeley, along with his student, Hartland Snyder. They realized that stellar collapse could create not only neutron stars but objects even more dense—so dense that light *could not overcome their gravitational attraction and escape.* Not long after this work, Oppenheimer threw himself into the Manhattan Project (the US project during World War II to develop a nuclear weapon) and subsequently into scientific administration and science policy, and never returned to this observation, which many consider his most significant achievement in pure science. It was John Archibald Wheeler, a Princeton physicist, who would eventually analyze the consequences of the Oppenheimer-Snyder work. Indeed, it was Wheeler who coined the term *black hole* to describe these objects. What Wheeler and subsequent investigators understood is that if the clump of material left behind in the exploding star is heavy enough, it distorts space and time in such a way that it disappears forever from view, just as we can no longer see a ship at sea once it passes beyond the horizon, due to the curvature of the earth. (When we look out upon the ocean, the distance to the horizon is about 22 miles if viewed from an altitude 100 meters [330 feet] above sea level.) The black hole is characterized only by a few properties—its mass, its electric charge if it has one, and the rate at which it spins around its axis (as the earth spins around an axis in very nearly 24 hours). Every other bit of information about the original star appears to be lost.

This horizon of the black hole—usually called the event horizon—is a strange place. It can be thought of as the surface of the black hole. At this distance from the center, time and space get mixed up—time acts like space, space acts like time. This is, in fact, the point of no return for objects falling into the black hole, and the farthest reach for light that starts out in the interior. Newton's notion of absolute time has atrophied even further.

For a big enough black hole, however, travelers passing through the horizon in a rocket would not notice anything. It would only be in their inability to communicate with mission control that they would perceive a problem. The real catastrophe would come as they approached the center of the black hole, where, like our neutron star traveler, they would be ripped apart. By this point, it is not clear that our notions of space and time make sense at all (the poor space traveler won't care). Physicists and mathematicians refer to the center of the black hole as a *singularity*. At this point (and close to it) Einstein's equations no longer make sense. What really happens at the center?

When I was a graduate student, black holes were still the subject of speculation. Astronomers had one candidate, an object called Cygnus X-1, a system of a visible star and another, compact object rotating around each other, about 6,000 light-years away from earth. The second object was massive and detectable through the effects of its motion on the first. The system also emitted X-rays. Over time, Cygnus was confirmed to contain a black hole, by studying the radiation emitted by the pair (not the black hole itself, of course). The X-rays are emitted by the debris of the star as it is sucked in by the black hole. The mass of the black hole can be inferred from its effects on the motion of the visible star. Indeed, astronomers *know* that any object this massive, if formed from the collapse of an ordinary star, is a black hole.

By now, black holes are almost a commonplace. There are many known similar to Cygnus X-1, paired with other stars and seen by their X-ray emission. More massive black holes have been discovered in the last few years through the emission of gravitational waves as they collide with one another. Perhaps even more dramatic are the supermassive black holes that have been discovered at the centers of many galaxies, including our own (whose mass is about 4 million times the mass of the sun). So nature confronts us directly with the disruption of space and time called for by Einstein's theory.

In 2020, the Nobel Prize was awarded to Reinhard Genzel, Andrea Ghez, and Roger Penrose for their work on black holes. Ghez studied the supermassive black holes and I was particularly happy to see her included. In the early 2000s, I was invited to speak at a conference about the future of particle astrophysics, sponsored by the National Science Foundation and the Department of Energy. In those days, PowerPoint software was just becoming popular. Instead, I prepared a talk using markers on plastic sheets (transparencies, as they were called). In my session, the first speaker was a high-level NASA administrator. He gave a talk using PowerPoint with gorgeous slides, on the agency's plans to image the black hole at the center of our galaxy, then a subject of speculation. I was depressed about my old-fashioned technology. Fortunately for me, the next speaker was a famous telescope designer, Roger Angel, whose transparencies were all in black and barely legible. At least mine had multiple colors and were neat. But the real lesson of the story is that, for all the slick presentation, barely two years later, I learned that Andrea Ghez had unveiled the black hole by studying trajectories of stars orbiting around it, and I was using visuals from her website in my classes. Ghez will appear later in our story.

The Big Bang

When we contemplate the universe as a whole, our notions of space and time come under even more extreme threat. Time seems to have a beginning.

In Einstein's theory, it is energy which bends space-time. But even at the surface of the sun, or at Mercury's orbit, the resulting effects are very tiny. So Einstein and others bravely turned to the largest amount of energy they could contemplate—the universe as a whole. I say bravely because, at that time, compared to what is now known, astronomers had limited understanding of our own Milky Way galaxy, much less the detailed maps we have out to 13 billion light-years today.

Even with the limited knowledge of the early twentieth century, putting the detailed features of the universe into Einstein's equations would have been a problem inaccessible to the paper and pencil techniques of the age. To get started, Einstein adopted a simple picture of the universe, one which, at first sight, seems rather crazy. He assumed that, anywhere you look, and in whatever direction you look, the universe appears the same. He wasn't really being totally crazy—he was adopting as a hypothesis that, viewed on very large scales or viewed in a very coarse way, the universe had these properties. Think of viewing the earth from space. With your eye, it's a sort of painted sphere; you can't resolve all the detailed structure on the surface. This was the nature of Einstein's "Cosmological Principle." But in those days, there was no evidence for this hypothesis, even in this crude form.

The outcome of this assumption was something entirely new—a model of the whole universe, a model with precise predictions for experiments and observations! The relevant equations were first derived and solved by Alexander Friedmann, a Russian physicist working in Saint Petersburg, in 1922. The most remarkable feature of the solutions is that the universe is not *static*. It expands as time passes. This statement is rather puzzling. What does it mean for the universe to expand? An analogy, which is not far from the actual mathematics, is to consider blowing up a balloon. Before blowing it up, mark it with dots, representing stars or galaxies in the two-dimensional universe which is the surface of the balloon. Now what happens as you blow up the balloon? Well, as the balloon expands, the dots move farther and farther from each other. From the perspective of any one star (dot), the others all appear to be moving away. Einstein's theory predicted exactly this behavior—except in a world of three spatial dimensions rather than two. It predicted as well that the rate at which the stars move away from us would be proportional to their distance from us.

To appreciate the nature of this leap, it is important to understand how limited was human knowledge of the universe on large scales at this time. In fact, it was only in this period that astronomers, most notably Edwin Hubble, discovered the existence of galaxies other than our own. Hubble, born in 1889, took a winding path to astronomy. At the request of his father, he first studied law. He then spent a period teaching high school, entering graduate school in astronomy at the University of Chicago after his father died. Following a brief stint in the Army during World War I, and further studies of astronomy at Cambridge University in England, he took a position at the Mount Wilson Observatory in Pasadena, California, in 1919. This was then the home of the world's largest telescope, giving him as broad a view of the universe as one might hope. Many astronomers of the day believed the Milky Way *was* the universe. Hubble's research changed that. Having identified different galaxies, Hubble undertook to measure the motion of other galaxies relative to our own. He found that, on average, the galaxies were all moving away from us, at a speed proportional to their distance from us. That proportionality constant became known as the Hubble constant.

I remember first learning of Hubble's measurements in a colloquium when I was a graduate student. The speaker, the astronomer Virginia Trimble, paused, and then said this result might be interpreted as showing that Copernicus was wrong; we really are at the center of the universe. She then offered the alternative possibility—as we saw before, thinking of the universe as the surface of the balloon, it looks the same everywhere as it expands.

In any case, Hubble's initial results were exactly in line with the predictions of Einstein's theory. Hubble's measurements were actually not of terribly high quality. His result for the rate of expansion of the universe was off by almost a factor of 10 from the number we know today. But the work launched what has now been more than a century of activity understanding the history of the universe, probing its large-scale structure and testing Einstein's theory.

Starting with the universe as it is today, Einstein's theory predicted that the universe is expanding. Correspondingly, if we look back in time, the universe was contracting. If we look far enough back, the universe was infinitesimally small—things were packed closely together—closer together than one could possibly imagine. Mathematicians described the point at which time begins in Einstein's equations as a singularity. The equations cease to make sense. This is the moment we have come to call the *big bang.* Einstein's theory breaks down at this time. Exactly when this breakdown happens and what happened before will be central issues in our explorations, but for now, we'll view the result as did Einstein and his contemporaries, as a feature of the equations.

The notion that time has a beginning, depending on your general outlook, can be a troubling one or an appealing one. It certainly troubled the great British astronomer Fred Hoyle, who viewed it as almost quasi-religious and unscientific. It was Hoyle, in fact, who coined the name big bang during a radio interview, and he did not mean it as a compliment to the theory (he later claimed that he did not actually mean it as an insult). But over time, the evidence in support of the theory has become overwhelming, as we will see. At the same time, the name has stuck. In 1993 the magazine *Sky & Telescope* ran a contest to establish a new name, reflecting the theory's improved status. But by then, big bang was so popular that they realized they had better leave well enough alone.

A Brief History of the Universe

As we look to the distant past, our conception of time, as revised and refined twice by Einstein, is safe at least until a small fraction of a second after the big bang. So we can try to construct a history of the universe back to extremely early times. We have reliable and well-understood evidence about what the universe was like three minutes after the big bang. We can trace various eras after that, and can look at least tens of billions of years into the future.

As we look back in time, the objects we see—stars, galaxies, particles of dust—all squeeze together. At early enough times, planets, stars, galaxies, break up into collections of atoms. As they smash together, they become hot. As the universe (in this backward video of its history) becomes still smaller, the universe becomes still hotter.

It's dizzying to run the clock backward this way. A more sensible approach was developed by George Gamow, a Russian émigré physicist, and his graduate student, Ralph Alpher, shortly after the Second World War. Gamow escaped Stalinist repression in the Soviet Union, coming first to France in 1933 and then to the United States the following year. During the remainder of his career, he served on the faculties of Washington University in Saint Louis and the University of Colorado in Boulder. He made notable contributions to nuclear physics. He was a successful writer of popular science, and something of a prankster. But particularly important were his steps to take Einstein's cosmology to times before the formation of stars and planets. With this came a plausible history of the universe from extremely early times, and the possibility of observational tests well beyond the expansion observed by Hubble.

Gamow and Alpher started by assuming that there was a moment, just seconds after the big bang, when the universe was *extremely* hot. The temperatures they contemplated were millions of times higher than those at the center of the sun, about 27 million degrees Fahrenheit (about 45 million degrees centigrade).

Temperature is a measure of the speed at which atoms and molecules move. The molecules in the air around us move fast—at room temperature, typically about 100 meters per second, or something in the vicinity of 100 kilometers per hour. They move in all directions, frequently colliding with each other (and with us, which is what gives us the sense of warmth).

The molecules are moving fast, yet extremely slowly compared to the speed of light, after Einstein the gold standard for speed. In the core of the sun, where the temperatures are 10^5 times, or 100,000 times, room temperature, atoms are moving at about 10^{-4} times the speed of light. This is fast enough that electrons are not able to stick to protons; the core of the sun is mostly ionized gas. But Gamow and Alpher contemplated temperatures where not only was all the gas ionized but atomic nuclei themselves could not hold together. They pictured a universe consisting of protons, neutrons, electrons, and photons (the particles that make up light, and which will be important players in almost all our stories) moving separately and almost freely. They imagined temperatures of about 10^{14} degrees (it doesn't much matter whether this temperature is Fahrenheit, centigrade, or Kelvin; a trillion degrees Fahrenheit, for example, is about two trillion degrees centigrade, and only 273 more degrees Kelvin).

In addition to the protons, neutrons, electrons, and photons at these high temperatures, there is a bizarre particle, which will appear repeatedly in our story, called the neutrino. Neutrinos can keep the number of neutrons and protons essentially the same. They do this by playing a dual role. First, colliding with protons, they can produce neutrons and other particles. Second, neutrons can undergo radioactive decay, producing neutrinos, protons, and electrons. Photons collide with protons, keeping protons and neutrons from joining up to form more complicated atomic nuclei and, with electrons, preventing the formation of atoms.

But as the universe expands, it cools, and its various constituents move more slowly. Eventually, the neutrinos don't have enough energy to turn protons into neutrons. Some of the neutrons decay; some join with protons to form an isotope of hydrogen, where the nucleus has one proton and one neutron, called deuterium. Protons and neutrons can also join to form the nuclei of more complicated atoms, such as helium and lithium. In fact, much of the matter of the universe, for nuclei such as deuterium, helium, and lithium, is formed this way. Heavier nuclei are mainly formed much later, from hydrogen burning in stars. The process in which the nuclei of the lighter elements are formed in the early universe is known as primordial nucleosynthesis. It takes place roughly 3 minutes after the big bang. For some time, we've known enough about nuclear reactions and the nature of the universe to predict how much of each light element there should be. Astronomers measure the fraction of the universe in hydrogen, helium, and the other light elements, and these agree pretty well with the predictions of the theory.

But another dramatic prediction emerges from these considerations. The universe, at this stage, is still extremely hot. It is not cool enough to form atoms until the temperature is a little below a million degrees Kelvin, or about twice that in Fahrenheit. At the time Gamow and Alpher did their work, they knew only roughly how old the universe was at this moment. We now know that neutral atoms formed at about the 100,000-year (10^5 year) mark after the big bang. Just as the abundances of the light elements are a relic of the first three minutes, there is an even more distinct fossil of this 100,000-year mark—a bath of microwaves that fills the universe, the cosmic microwave background radiation (CMBR). This time is known as the *cosmic recombination* time.

Before this moment, photons—the particles that make up light proposed by Einstein in 1907—collide constantly with electrons and protons, traveling only tiny distances between collisions. Once the universe consists of neutral atoms, though, the photons can pass almost completely unimpeded through the universe. Gamow and Alpher realized that these photons would be around today. At the time of cosmic recombination, they would have wavelengths similar to that of visible light. But due to Einstein's redshift, by now the typical photon would have a much larger wavelength, like that of the microwaves in your oven.

The CMBR was discovered in a spectacular and serendipitous way in 1964, almost two decades after Gamow and Alpher's work. In those days, there were several great industrial laboratories belonging to major corporations. Particularly prominent were the Bell Labs, at several sites in New Jersey. These were run by AT&T, which then enjoyed a monopoly on telephone communication. Another laboratory was IBM's research center in Yorktown Heights, New York. Scientists at these labs engaged in work directly relevant to their corporate bottom lines, but they also often had freedom to pursue questions they found scientifically interesting. Two physicists, Arno Penzias and Robert Wilson, working at Bell Labs in Holmdel, New Jersey, built a large antenna designed for radio astronomy. Good experimentalists as they were, they first wanted to test their instrument, and they trained it on a part of the sky in which they expected no signal, aiming to *check* that they saw nothing. Instead, they saw something dramatic, somewhat similar to the background static you may hear as you tune your car radio. They first thought this reflected a problem with their instrument, so they began a series of tests and checks to find the source of the problem. They were so skeptical of the signal that when they noticed that pigeons had nested in the antenna, they speculated that bird dung was the source of the problem. They dismantled and cleaned the antenna. Still the signal persisted. Finally, they met with astrophysicists Robert Dicke and Jim Peebles of Princeton. Both had been working hard on the problem of the CMBR, Peebles on the theory and Dicke on developing an experiment to look for it. They explained to the two Bell Labs researchers the expected frequency and strength of the radiation, and Penzias and Wilson immediately set to work to determine if this was the signal they had discovered. Of course, the point to this story is that it was. The initial data was meager—the strength of the signal was known only for a few values of the frequency, but over the next few years, dedicated measurements drastically improved the situation. It soon became clear that the strength of the signal, and the manner in which it depended on the frequency, was exactly as Gamow and Alpher, and subsequently Peebles and others, had predicted. By now, in fact, the agreement between theory and experiment is about as close to perfect as it can be.

The spectrum of the microwave radiation depends on the temperature of the radiation in a universal way. So the measurement of the radiation provides a measurement of the temperature of the universe today. But it provided something more: a test of the cosmological principle. Studying the radiation arriving on earth from different directions, astrophysicists found that the temperature is the same in all directions, to a high degree of accuracy. Since the radiation comes from vast distances (it's been traveling toward us for 13.5 billion years), this is evidence that the universe, on very large scales of distance, is the same everywhere and in all directions, as Einstein had postulated. In parallel, over the last few decades, surveys of very distant galaxies have established that matter is distributed uniformly and in the same way in all directions when viewed on large scales. Astronomers say that the universe is homogenous (smooth) and isotropic (the same in all directions).

But in a sense, this was too good to be true. The universe is certainly not perfectly homogeneous and isotropic. The most reasonable explanation for this seemed to be that, at the beginning, it was a little bit inhomogeneous and anisotropic, and that these imperfections grew as the universe expanded and aged. But for a long time, there was no evidence for this in the CMBR. In fact, the CMBR was soon measured to be homogeneous and isotropic to a part in 10,000.

So we have a history of the universe, with experimental and observational support, from times about three minutes after the big bang, to the present, more than 13 billion years later. Yet we still don't know what the singularity of Einstein's equations means, and whether the universe truly had a beginning.

STEP TWO

CHAPTER 4

CAN QUANTUM MECHANICS PREDICT THE FUTURE?

In this chapter, we head away from the cosmically huge in the opposite direction, to explore the realm of the very small. Our understanding of nature is about to alter in profound ways. When I was a student, I had a girlfriend who complained about my interest in physics—"It's so mechanistic." Newton's view of the universe *was* very mechanistic. Tell me where a planet is now, and how fast it's moving, and with a good computer I can tell you where it will be at any later time. Maybe it's just me, but I don't really find that so boring. Still, I was somewhat embarrassed, and I tried to impress with an interest in history and literature.

Einstein's relativity theories make the description of motion much more interesting. Yet it remains true that if an astronomer knows the position and speed of a galaxy now, she can determine, using Einstein's equations, where it will be at any later time. But things look very different at the scale of atoms and things smaller. During those student days, I learned that science has had to replace the Newtonian worldview with something quite strange. Nature is described by mysterious rules that don't align with any intuition we have for the way objects move, and the fact that humans figured this out was—and remains—for me totally awesome. This is the world of the quantum.

While I am not an evolutionary biologist, the sort of predicting of the future that Newton developed would seem a rather natural extension of our evolutionary programming. To pursue food, avoid falling debris, aim a projectile, or face other challenges to our survival, short-term prediction of how things fall and move through the air is essential. Yet when, in the late nineteenth and early twentieth centuries, physicists began seriously to study atoms, they found to their consternation that the rules of Newton and Einstein did not apply. This perhaps should not be a surprise: There was no evolutionary imperative that required humans be able to understand nature on scales of atoms and things even smaller. That we should be able to figure out working principles is not a given. But humans did.

Speculation about the constituents of matter dates back to at least the ancient Greeks, but for centuries this was nothing *but* speculation; these were ideas about how the world *might* work, based on zero evidence. In the nineteenth century, all of this changed dramatically. Chemists noticed patterns among the elements, suggestive that the samples in their test tubes were collections of large numbers of atoms with characteristic properties. Michael Faraday, who we encountered in our discussion of electricity and magnetism, performed experiments with electric currents, which he realized could be explained if there were atoms, themselves consisting of some sort of particle with charge. The state of the atomic hypothesis is reflected in his statement that "yet we cannot resist forming some idea of a small particle, which represents it to the mind . . . there is an immensity of facts which justify us in believing that the atoms of matter are in some way endowed or associated with electrical powers, to which they owe their striking qualities." James Clerk Maxwell, who also contributed so much to our understanding of electricity and magnetism, developed a theory in which atoms accounted for the observed properties of gases. Like the Greeks (the word *atom* is Greek for "uncuttable"), he believed that atoms were unbreakable: "Though in the course of ages, catastrophes have occurred and may yet occur in the heavens, though ancient systems may be dissolved and new systems evolved out of their ruins, the [atoms] out of which these systems are built—the foundation stones of the material universe—remain unbroken and unworn. They continue to this day as they were created—perfect in number and measure and weight . . ." Maxwell's successful picture of gases containing large numbers of atoms, moving according to Newton's laws, still posed no threat to the classical worldview. But there was not yet a precise picture of the atom.

The real challenges came as the nineteenth century ended and the twentieth dawned. Two in particular were poised to doom Newton's worldview. The first was the discovery of radioactivity, in 1895, by Henri Becquerel. Becquerel was the third of what would be four generations of his family who would hold a chair of physics at the Museum of Natural History in

Paris. While conducting experiments meant to understand the recently discovered phenomenon of X-rays, he found, almost by accident, that uranium gives off rays capable of leaving an image on a photographic plate—light-sensitive material used to make pictures long before the era of digital cameras, much less cell phones.

Marie Curie, then a young scientist eager to find interesting problems to work on, decided to study these "Becquerel rays." She quickly demonstrated that the radiation was a property intrinsic to the atom: The same element, combined in different compounds, still radiated in the same way. She also recognized that there was more than one element responsible for the radiation in her—not so pure—uranium samples. From the perspective of our story, she made one other, critical breakthrough quickly: She demonstrated that the amount of radiation was proportional to the amount of the particular radiating element. This, she and others realized, means that one can't predict when an individual atom is going to decay, but can determine only that there is a probability that any given atom will decay, say, in the next second. Newton's overthrow was imminent.

It is worth pausing to appreciate how impressive Marie Curie was as a scientist, and how remarkable were many aspects of her life. Born in Poland (her name at birth was Maria Sklodowska), she first helped support her older sister to come to Paris to study medicine. Her sister, Bronislawa, returned the favor, helping to support Marie as she studied physics at the Sorbonne, beginning in 1891. The research I've described was part of her work toward her PhD. In this period, she met and married another, slightly older and already well-established physicist, Pierre Curie. When Marie decided to pursue the search for the other radioactive elements in her samples, Pierre decided that this was more important than his own work at that time and dropped it to serve as her assistant. One scientist who worked for a period in their lab wrote, "There have been and there are scientific couples who collaborate with great distinction, but there has not been a second union of woman and man who represented, both in their own right, a great scientist. Nor would it be possible to find a more distinguished instance where husband and wife with all their mutual admiration and devotion preserved so completely independence of character, in life as well as in science." She would go on to win two Nobel Prizes, one in physics and one in chemistry, and to be the first female professor at the Sorbonne. All this is in an era before women even had the vote!

Another physicist who, as we'll see, played a crucial role in understanding the atom was Ernest Rutherford. He also threw himself into the study of radioactivity after Becquerel's discovery. Originally from New Zealand, he spent time as a professor at McGill University in Montreal and then at the University of Manchester, before taking a position at Cambridge University. Rutherford was a devoted son, and later a devoted husband and father. In 1902, he wrote a letter to his mother in New Zealand, which convinces me of Curie's brilliance: "I have to keep going as there are always people on my track. I have to publish my present work as rapidly as possible in order to keep in the race. The best sprinters in this road of investigation are Becquerel and the Curies in Paris, who have done a great deal of very important work on the subject of radioactive bodies during the last few years."

At about the same time as the discovery of radioactivity, the second big challenge arose for classical physics: understanding the nature of the radiation emitted by objects at high temperature. If you stand near a hot object, say a hot pan, you feel the heat. This is not just because the pan warms the molecules of the air but also because it gives off a type of electromagnetic radiation: infrared radiation. This radiation is not visible to the eye, but it causes vibrations in the molecules in your skin, which you experience as warmth. This sort of radiation was studied experimentally in the late nineteenth century and is known as *blackbody radiation*. The whole universe acts as a blackbody; with its current temperature, 2.7 degrees Kelvin, this is the cosmic microwave background radiation that, we have seen, figures crucially in our present understanding of the big bang. Following Maxwell's ideas for the behavior of systems of particles with a temperature, one could use classical physics to predict how much radiation of a given frequency a blackbody would produce. This predicted that a hot object gives off not just infrared radiation but radiation of every type: microwave radiation, infrared radiation, visible light, X-rays, and gamma rays. In fact, there are an infinity of possible types of radiation, and correspondingly, the theory predicted that the hot object gave off an infinite amount of radiation energy. This clearly made no sense, and this democracy among frequencies did not agree with the experimental observations, where the radiation from an object, say at 300 degrees centigrade, was mostly in the infrared.

Enter Max Planck. Planck was, in many ways, a very traditional, conservative individual and scientist, but to resolve this puzzle, he took an extremely radical step. In Newton's physics, energies can take any value; they form what mathematicians call a continuum. In other words, there can be 1.000000 units of energy, or 1.000001 units of energy; the energy can vary to as many decimals as one wishes. Planck made the hypothesis that for atomic systems, energies can take only particular, discrete values, say 1.00000, 2.000000, etc. These values, he proposed, are much larger for visible light than for infrared light, and much larger again for X-rays, and still larger for gamma rays. For the blackbody radiation, the visible light, X-rays, and gamma rays are not produced because the temperature of the body isn't as high as the minimum needed to produce the lowest energy packet—or *quantum* of energy. Planck's hypothesis agreed beautifully with the data. In 1905, in the work that would win him the Nobel Prize, Einstein applied Planck's hypothesis in the other direction, asserting that the *absorption* of light by matter, as well as its emission, takes place in discrete units of energy. In 1907, he took the next step, saying that light itself comes in discrete quanta, or particles, called *photons*. The Newtonian view was now in deep trouble. Nothing in Newtonian physics could account for this discreteness of energy, what would come to be called *quantization*.

The death blow for the classical worldview was provided by the discovery of the structure of the atom. The end of the nineteenth century and the beginning of the twentieth saw the development of tools that allowed one to "see" the atom and its working parts. The first step was the discovery, by J. J.

Thomson of Cambridge University, of the electron. In 1897, Thomson was studying the flow of electricity in gases. He was able to show that this electricity was actually the flow of large numbers of a small particle—the electron—and measure its mass and electric charge. He could also demonstrate that the electron was much lighter than a hydrogen atom.

So atoms were no longer unbreakable, but what were they? Thompson put forth a picture in which the electrons sat in a sort of positively charged goop, so that overall the atoms were electrically neutral. The electrons could be pulled out of the atom, leaving behind other electrons and the goop—the charged ions of such interest to chemists. Thomson was British, and this model of the atom was called the "plum pudding model." Exactly how this structure was held together—and for that matter, what it really was—was unclear, but Thomson and his contemporaries still thought about the problem in a Newtonian framework.

This model didn't last.

Another new tool to explore the atomic world came to hand. The great British experimentalist Ernest Rutherford realized that the fast particles emitted by radioactive materials could be used to probe the structure of atoms. He was able to direct the particles from a source of radioactivity toward a thin film of gold, and by looking how the particles scattered from the gold (again using photographic plates), he was able to map out the structure of gold atoms. If Thomson's model was correct, these particles would pass through the goop almost undisturbed. Instead, he found they frequently reversed course, bouncing back off the atoms. This came as a complete shock. Of the experience he wrote, "It was quite the most incredible event that ever happened to me in my life. it was almost as incredible as if you fired a 15-inch shell at a piece of tissue paper and it came back and hit you." His observations demolished Thomson's model.

Rutherford did much more. He identified the particles that made up the different known types of radioactivity. These were of three types. Gamma rays were composed of high-energy particles of light, alpha rays consisted of nuclei of helium atoms, and beta rays were electrons. It was the charged alpha particles which were his tool for uncovering the structure of the atom. Rutherford worked out the theory of how these particles would scatter off the nucleus, and it was the comparison of his measurements and the theory which nailed down the size of the different parts of the atom. While each of these discoveries would constitute an extraordinary scientific career, he did still more. It was Rutherford who discovered the proton; his colleague (and former student) James Chadwick, pursuing Rutherford's suggestions, would later discover the neutron. Rutherford also provided a good measurement of the size of atoms—about 10^{-8} cm, and the size of the nucleus, 100,000 times smaller, or about 10^{-13} cm across. Radioactivity had provided the tools to explore about eight powers of 10 smaller than was possible with light microscopes.

Niels Bohr realized the extraordinary implications of Rutherford's discoveries. Then twenty-seven years old, he came shortly after the discovery of the nucleus to work as an assistant—what we would now probably call a postdoc—in Rutherford's lab. Born in Denmark and educated in Copenhagen, he would play a dominant role in the quantum revolution.

He recognized that Rutherford's results raised huge puzzles. First, if Newton's (or Einstein's) picture was right, there was no way to understand why atoms were identical. Their chemistry would depend on exactly where each electron was in the atom at a given moment and how fast it was moving, and all of this would depend on the history of the atom. In other words, the chemistry of every individual atom would be different. This problem could be highlighted another way. If electrons were governed by Maxwell's theory, as they orbited the nucleus, they would emit light. Light carries energy, so the electrons would gradually slow down and fall into the nucleus. Atoms of the sort that Rutherford had discovered would not exist at all.

Bohr was well aware of Planck's work on discrete quanta of energy, and Einstein's photon concept. To address the puzzle of the atom, he took their ideas a step further, completely and finally upending the rules of classical physics. His proposal asserted that the electrons could move only in orbits, or states, with definite speed, and correspondingly a definite energy. A higher energy state could "jump" to a lower energy state, emitting a photon to carry off the extra energy. Or a lower energy state could absorb a photon, jumping to a higher orbit. By fiat, he resolved both puzzles. The lowest energy state was simply stable. It didn't decay. Every atom in its state of lowest possible energy was identical to every other such atom. His predictions for the energies of the emitted photons agreed with known results for the spectrum of light emitted by hydrogen, a spectacular success. But Bohr's model was very specific to hydrogen and couldn't readily be generalized to describe other types of atoms. For almost a decade, Bohr and others tried to extend these rules, with only limited success. The rules they proposed were simply too ad hoc. Newtonian physics was overthrown, but what was the framework that might replace it?

A critical step was taken by the French physicist Louis de Broglie. Comparing Bohr's postulate with Planck's, he made the seemingly outlandish proposal that the electron was a wave, like light, rather than a particle. In this way, Bohr's rules were just an application of Planck's. A few years later, experiments demonstrated that, in some circumstances, electrons *do* behave like waves. Particles, waves—which were they?

A radically different framework for physics began to emerge in the early 1920s, when Werner Heisenberg and Erwin Schrodinger each put forward a new setting for the laws of nature. Heisenberg wrote down equations for Planck and Bohr's discrete states, which predicted how they change over time. Schrodinger took De Broglie's ideas to the next level, assumed that the electron might be described by a wave, and wrote an equation, similar in spirit to Maxwell's equations, which explained how the waves changed in time. Both equations gave Bohr's results for the hydrogen atom, without additional assumptions or ad hoc rules. They both accommodated more complicated atoms, like helium. Being of a form with which physicists were already familiar, Schrodinger's setup at first had greater appeal, and the crucial equation of quantum mechanics has come to be known as the Schrodinger equation. But in the search for the meaning of it all, as we'll see, Schrodinger and De Broglie quickly fell far behind, while Heisenberg, Bohr, and two other theorists, Paul Dirac and Max Born, put together a complete— if very weird and in some ways troubling—picture.

Paul Dirac was an English physicist who spent most of his career at Cambridge University, moving to Florida State University late in life. He has been called "The Strangest Man." Socially, he was indeed quite awkward, but he played a crucial role in the development of quantum physics. It was Dirac who understood that Heisenberg's and Schrodinger's formulations of the quantum theory were really the same and who put together a presentation that propelled the application of quantum mechanics to a broad range of phenomena: completely understanding the periodic table in terms of properties of atoms as well as the physics and chemistry of molecules. Other forms of matter—solids, liquids, and gases—could be readily understood, and quantum mechanics now provided a setting to speculate about phenomena on much smaller scales.

But it was Max Born who was the first to figure out just how the quantum mechanics of Heisenberg and Schrodinger superseded the science of Newton. Born, a professor at Gottingen in Germany, was trying to understand processes like those in Rutherford's experiments, using the new mechanics. If Newton's laws describe particles like the electrons or protons, then, starting with knowledge of the location and speed of all the particles, they predict the outcomes of measurement of the location and at any later time. Born realized that in quantum mechanics, Schrodinger's and Heisenberg's equations predict the *probabilities* of outcomes of measurement. Schrodinger's wave function associated a number with every point in space, whose *square* was the probability to find the particle at that point. Other such numbers give the probabilities of radioactive decay at a given instant, and the probability of other outcomes of experiment. Quite generally, the notion that one can predict outcomes with certainty was replaced by the prediction of such probabilities.

Born thus made precise the limitations on our ability to predict the future that Curie, Rutherford, and others had already inferred from the study of radioactive decay. Werner Heisenberg, with his *uncertainty principle*, framed the limitation of our knowledge, in sharp terms. He considered a hypothetical set of measurements of the motion of an electron, and realized that the more precisely one measured the position of an electron, the less well one knew its speed, and similarly the better one measured the speed, the less well one could measure its position. This limitation was fundamental, something built into the laws of quantum mechanics. One could not get around it by improving the measuring equipment. The principle applied not just to electrons but to all particles, and not just to position and velocity but to almost anything about a system one might hope to measure.

In one sense, what the quantum mechanicians did to Newton is not so different from what Einstein did. They altered the questions and the way we answer them, but for most of our day-to-day experience, it doesn't matter. Nor does quantum mechanics give up on Newton's ultimate goals. It still provides an understanding of physical systems and predicts their behaviors. In fact, quantum mechanics makes incredibly sharp and precise predictions. It does, however, change the nature of the questions one asks. Instead of solving Newton's equations to find where a particle is at different times, one solves the Schrodinger equation to find the probabilities of different events in the future, given knowledge of the wave function at some time in the past. With this, one can find the discrete energies that Planck and Bohr postulated. But one can do much more. Bohr also postulated rules for how atoms would absorb and emit light quanta—photons. With Schrodinger's theory, these rules were automatic, and one could predict much more: the actual probabilities that an atom would absorb a passing light quantum or that, having absorbed one, it would emit another. As we'll discuss further, these predictions are incredibly, even ridiculously, successful. The quantum mechanicians could also study atoms much more complicated than hydrogen, with more electrons, and they could work out much finer predictions about the structure of hydrogen itself. The Niels Bohr Institute at the University of Copenhagen (funded by the Carlsberg Foundation, a charitable arm of the brewing company) was the center for the synthesis of these ideas, and the understanding of quantum mechanics that emerged in this period has come to be known as the *Copenhagen interpretation*.

Still, at the level of atoms and things smaller, quantum mechanics teaches us that the intuitions we have about nature from our day-to-day experience are just wrong. It amazes me that human beings figured all of this out. The evolutionary imperative to understand the trajectory of a falling object, or a thrown stick or spear, is reasonably clear, but our survival was not, until quite recently, dependent on our understanding of how atoms work. Bohr, Heisenberg, and Born, who did so much to put this picture together, at times along the way despaired, worrying that things happening on the atomic scale might simply be beyond human powers of understanding. But with the success of quantum mechanics in explaining atoms, including many features of the periodic table, and the absorption and emission of light, opened new vistas on natural phenomena. One could try to answer previously undreamed-of questions.

With the overthrow of Newton's worldview came, as well, the overthrow of Einstein's. Einstein, with his photon concept and other work that generalized Planck's radiation law, did much to set the stage for the quantum theory. Einstein also laid the groundwork for studying the quantum mechanics of systems of many atoms, with a young Indian theorist, Satyendra Nath Bose. He did work that paved the way for understanding emission and absorption of radiation by atoms, which would be crucial, much later, to the development of the laser. But as quantum mechanics matured in the 1920s, Einstein became more and more uneasy about it. He had long debates with Bohr, in which he tried to poke holes in the probability interpretation. He did not succeed.

Reconciling Einstein's special relativity and quantum mechanics would require further radical steps. Schrodinger's and Heisenberg's equations, while quite successfully reproducing many atomic phenomena, did not respect Einstein's principles, in much the same way that Newton's did not. This was recognized early on, but the first attempts to find equations compatible with special relativity ran into serious difficulties.

A crucial step was taken by Dirac. He made an inspired guess as to an equation that would describe the motion of a relativistic electron. This equation built in the known features of the electron (in particular a property called spin) from the beginning and accounted for many fine details of atomic structure. But the equation and its interpretation still posed serious challenges. The electron seemed to be unstable. Dirac, after some missteps, had another remarkable insight: He realized that the equation predicts the existence of antimatter—in this case a particle exactly like the electron in every way but with the opposite electric charge. With this rethinking of his equation, Dirac eliminated the problems. This was a first. In Newton's world, and even in Schrodinger's nonrelativistic universe, one must simply assert the existence of a particle of a particular type. Now the theory, by its very nature, predicted that the electron is accompanied by another particle with very definite properties. This particle was dubbed the positron, in deference to its positive charge. Dirac's theory predicted that an electron and a positron, if they meet, can annihilate, producing other forms of energy (high-energy photons in this case).

Dirac made his prediction in May 1931. The positron was actually discovered a few months later by Carl D. Anderson. Anderson was a brilliant experimentalist, and he was skeptical of theory and theorists. He made his discovery shortly after receiving his PhD, with an instrument he built to study cosmic rays, the energetic radiation (particles) from space that constantly strikes the earth. Anderson was dismissive of the importance of Dirac's theory for his discovery, and clearly annoyed that Dirac's paper had appeared just a few months before: "Yes, I knew about the Dirac theory . . . But I was not familiar in detail with Dirac's work. I was too busy operating this piece of equipment to have much time to read his papers . . . [Their] highly esoteric character was apparently not in tune with most of the scientific thinking of the day . . . The discovery of the positron was wholly accidental."

Dirac's equation was remarkable in another way. Electrons were known, from a variety of experiments, to act as small magnets. This magnetic property was associated with the property of spin. The spin points in a particular direction, and this in turn determines the directionality of the magnet (much as a bar magnet can be oriented in one direction or another). Unlike position and momentum within the quantum theory as it had developed up to that time, the spin didn't have a nice correspondence to any classical notion. It was completely ad hoc. But spin was intrinsic to Dirac's theory. More striking, his theory got the connection between spin and the magnetism of the electron exactly right.

The Quantization of the Electromagnetic Field

With Dirac's equation and his discovery of antimatter, part of the problem of reconciling quantum mechanics and special relativity was solved. The charged particles of electrodynamics could now be considered understood. But what of the fields, the electromagnetic waves that make up light, microwaves, X-rays, gamma rays, and the rest of the electromagnetic spectrum? Just as Planck's and Bohr's quantized energies were accounted for by Schrodinger's equation, one should be able to understand Einstein's photons, discrete particles of light, in terms of such equations. The answer turns out to be that the correct equations are Maxwell's, but they have to be interpreted according to the rules of quantum mechanics. The fields themselves are *quantized.*

The work of Schrodinger and Heisenberg codified the notion that electrons have a particle aspect, as demonstrated by J. J. Thomson's experiments, and a wave aspect, as hypothesized by De Broglie. But photons posed more serious challenges. The early practitioners of quantum mechanics could understand photons moving through space and isolated from matter. They could understand the forces that held atoms together in terms of the *exchange of photons.* But they could only roughly estimate the rates for emission and absorption of photons by atoms. Meanwhile, the experimental measurements of these processes became more and more accurate. For theorists, there were, in fact, seemingly insuperable obstacles to precise calculations. The rules of quantum mechanics, when taken to hold rigidly, gave nonsense answers. The problem arose as one attempted to make approximate calculations of quantities, such as rates for an atom to transition between one energy state and another. There seemed a natural way to do a first approximation, followed by more and more accurate calculations. But as one proceeded in this way, at some point the expressions one thought should be written for these accurate calculations made no sense.

The difficulties were related to the uncertainty principle. If one calculated, for example, the electron mass, one had to allow the possibility that for a brief time the electron could turn into an electron and a photon. This violates conservation of energy, but Heisenberg's principle permits it. The trouble comes because the electron and photon can have any energy—the energies can be arbitrarily large. Each of these "virtual states" contributes to the electron mass, and adding this infinity of contributions leads to an infinite result. The same applies to anything one wants to calculate in these theories—the rates for emission and absorption of light, the detailed properties of atoms.

This problem resisted all attempts at solution for nearly two decades. But in the years immediately after the Second World War, there were a series of breakthroughs. In the United States, the principal players were Hans Bethe, a German Jewish refugee at Cornell, and two younger theorists, Julian Schwinger, working at Harvard, and Richard Feynman. All three had been involved in research related to the war effort. Schwinger had spent those years at MIT, working on the development of radar and related technologies. Bethe and Feynman had worked on the atomic bomb project at Los Alamos. All had acquired, during their war work, a deep understanding of the classical theory of electricity and magnetism. There was a fourth crucial player, Sin-Itiro Tomonaga, a Japanese theorist who, amazingly, had made progress on these questions during the war. Not surprisingly, his work was not known in the United States until well after the war.

Sensitive experiments in these postwar years were a crucial driver. The early, crude calculations were not sufficiently accurate to account for the new experimental results. These experiments included more precise measurements of the energies of states of atoms and of the magnetic properties of the electron. Theoretical progress quickly followed. Hans Bethe was a scientist of extraordinary accomplishments. In the 1930s, he had laid the groundwork for understanding how nuclear reactions power stars—one of the most important elements of modern astronomy. He had made numerous contributions to our understanding of quantum mechanics and of nuclear physics. In 1947, he heard about extremely precise measurements by Willis Lamb, then at Columbia University, of the energies of the photons emitted by hydrogen. At a conference in Shelter Island, New York, these results, which did not agree with the Dirac theory of the electron, were a central topic of discussion. In Dirac's theory, the electromagnetic field was treated as a classical object; the quantum nature of the field—the photons—were not properly taken into account. Possible strategies were discussed at the conference. During the train ride back to Cornell, Bethe did a rough calculation that accounted for most of the discrepancy between Lamb's measurement and Dirac's result. But to get a sensible result, Bethe had to set aside what he understood to be the rules of quantum mechanics, and it was not clear how one could improve Bethe's guess so as to calculate with greater accuracy.

A crucial step to solving the problem lay in confronting something troubling about the very way Bethe and others formulated the quantum theory of electromagnetism. Their calculations all singled out time, as opposed to space. But in Einstein's special relativity, time and space are on a very similar footing. Stating the problem is relatively easy; resolving it required acts of genius. It was Schwinger, Feynman, and Tomonaga who made the crucial breakthroughs. Establishing a method that was at every stage compatible with Einstein's principle allowed both efficient calculations and a conceptual framework that addressed the troubling aspects of the theory. Tomonaga was isolated in wartime Japan, and it was Feynman and Schwinger whose work had the most immediate impact. These two had remarkably different styles, in life and in their work. Schwinger was something of a dandy. While he attended City College in New York, a school often referred to as the "Harvard of the proletariat," he fancied Cadillacs and tailored suits. Feynman was a less polished character, but a showman who during his Los Alamos years would perform almost magical mathematical tricks (with Bethe) and broke into laboratory safes; in subsequent years, he would be known for playing the bongos while frequenting trendy clubs. Their physics was a similar study in contrasts. Schwinger's mathematics was sophisticated, to the point that for most physicists it was difficult to follow. Feynman's was, again, less polished but more intuitive, accompanied by pictures and seemingly arbitrary but precise rules. J. Robert Oppenheimer, famed for his leadership of the atomic bomb project, was in this period the director of the Institute for Advanced Study in Princeton and in this role was an arbiter of what was to be taken seriously in theoretical physics. He strongly favored Schwinger's precise, very mathematical approach, and was dismissive of Feynman. Those of us who saw Feynman as the picture of self-confidence in his later years, flamboyant and iconoclastic, have a hard time imagining how devastating he found Oppenheimer's criticism.*

---

*Through the years, Feynman has had many admirers and has been the subject of numerous books and articles. His own books, both textbooks and those for lay readers, have quite a following. In recent years, however, his sexist—even possibly misogynist—behavior has also come under scrutiny.

In the end, it is Feynman's viewpoint that triumphed. This had much to do with Freeman Dyson, a British theoretical physicist who had come to the United States after the war. He had struck up a close friendship with Feynman. The two of them had traveled together by car across the country. Dyson appreciated both Schwinger's point of view and Feynman's and was determined to reconcile them—and then convince Oppenheimer of the value of both. Dyson's approach, to this day, dominates the way in which theorists and experimentalists understand the quantum nature of the electromagnetic field.

With these techniques, the problem of the infinities in quantum field theory was conquered, and precise calculations were undertaken. At our present moment in time, the magnetic properties of electrons—the "magnetic moment," are among the most precisely measured and calculated quantities in the natural world, and both agree—14 decimal places. In Dirac's theory, where the quantum mechanics of the photon is ignored, this number would be 2. It's measured value today can be expressed as: 2.002 319 304 361 82. Feynman, Schwinger, and Tomonaga would share the Nobel Prize in 1965; Lamb was awarded the prize for his experiments in 1955.

What's Bothersome about Quantum Mechanics?

While Bohr, Heisenberg, Dirac, and Born quickly got with the program of quantum mechanics, several of those who had played pivotal roles at earlier stages could not swallow the demise of the Newtonian notion of causality. Among these were De Broglie and Schrodinger, both of whom wanted to think of the electron and similar waves as describing the actual movement of some sort of substance. While it was easy to understand their unease, it was also easy to dismiss their objections, which Bohr, Heisenberg, and others did. But then, in 1935, Einstein, Boris Podolsky, and Nathan Rosen formulated a challenge to quantum mechanics. This paper, and the puzzle it proposed, is so famous—and troubling—that it has become known as the EPR paradox. They started by defining a notion of what they called a "complete description of reality," and then argued that quantum mechanics failed to provide one. Their objection centered around an aspect of the uncertainty principle, the incomplete knowledge that one has of quantum mechanical systems.

The problem that bothered EPR can be studied in a real system. Just as an electron, with its negative charge, can bind to a positively charged proton to make an atom, so an electron can bind to a positron, it's antiparticle, and make an atom. This system is actually created and studied in the lab. It's a lot like hydrogen, in fact, except in one striking feature: Eventually, the electron and positron, which are separated on average by the typical size of an atom, or about 100-millionth of an inch, will find each other and annihilate. When that happens, they produce photons—either two or three. Extremely precise measurements have been made of the discrete energies and the amount of time, on average, it takes for annihilation to occur. Quantum calculations, using the methods of Feynman, Schwinger, and Tomonaga, successfully account for the measured results. For example, for the average time it takes the electron to find the positron and annihilate, the theory gives 7.03996 millionths of a second (microseconds), with an uncertainty in the last decimal place. The experimental measurement is in excellent agreement.

Given this success of quantum mechanics, it would be surprising were it to fail. But we can appreciate what bothered EPR if we consider in a bit more detail the prediction of quantum mechanics for what happens when positronium decays to two photons. Just as electrons carry a spin, which can point either up or down, so the photon comes with a spin, which can point up or down. In the case of the photon, the spin is perpendicular to the direction in which it moves. The spin is related to what we call the polarization of light. According to quantum theory, if one measures the spin of one of the photons to be up, then the measurement of the spin of the other photon has to be down. If one measures the spin of the first photon to be down, the other yields up. The problem is that, until one does the measurement, one doesn't know the result, just the probability of one or the other outcome for each photon. So we could take our two photons and have two observers, each a light-year away in the opposite direction, measure the spin of the photon that comes near them. Until the photon arrives, they have no idea what each of them will measure. When the photon passes by and they *do* their measurement, they immediately know what their partner measured, even though she's two light-years away. This really bothered Einstein. But it turns out that an array of experiments has by now demonstrated that this property of quantum mechanics is real.

While the EPR puzzle was long viewed as an oddity of interest to die-hard skeptics, in the last few years, this subject has taken on new life. The two photons from the positronium decay are said to be in an *entangled* state. There is information stored in the system in a very subtle way. This sort of information storage actually holds promise for the construction of extremely powerful computers capable of storing vastly more information and doing far more elaborate calculations than conventional ones. There are many challenges to making such "quantum computers" a reality. Perhaps the biggest is that these systems must be well isolated from their environment, or the information leaks out.

It is hard to resist talking about Schrodinger's cat. This was a seeming paradox formulated by Schrodinger, which gives pause to anyone who thinks about quantum mechanics. Schrodinger imagined a situation where there was a cat in a box connected to a bottle of poison gas. Now suppose, again, we have two photons from positronium decay. The second one passes through a spin detector attached to the gas vial. If the second photon has spin up, it will trigger the release of the gas, and the cat will die. If the spin is down, the gas would remain in the bottle and the cat would survive. The measurement of the first photon spin has probability 1/2 of yielding spin down, probability 1/2 of yielding spin up. There is no definite result without the measurement. The cat is alive and dead at the same time. Worse, for animal lovers, the measurement can kill the cat (or not), and the result is instantaneous with the measurement.

What Schrodinger failed to understand is a variation on the problem of quantum computing. There is far more information in this problem than just that contained in the two photons. The cat has a huge number of atoms, all with spins and other properties, as does the gas bottle, and the photons and all these atoms are deeply entangled. All this is meant to convince you that EPR and Schrodinger's cat are not objections to quantum mechanics. But you're still entitled to find the properties of quantum mechanics very unsettling.

The Triumph of Quantum Mechanics

With these successes of quantum mechanics in understanding atoms and photons, which comprise a theory known as *quantum electrodynamics*, humanity had an understanding of nature at the atomic level, on scales of length and time orders of magnitude smaller than the scales known in Newton's day, or even in Maxwell's. From the start of these developments immediately after the Second World War, this would, over the latter half of the twentieth century and the beginning of the twenty-first, extend to a complete description of length scales still eight powers of 10 smaller. The underlying laws governing the strong nuclear force and the weak force, a sort of cosmic alchemist that allows the creation of elements in the universe heavier than hydrogen, would be understood using the same principles of quantum mechanics and special relativity as governed quantum electrodynamics, in terms of a generalization of that theory known as the *Standard Model*. It is an extraordinarily successful marriage of theoretical and experimental physics.

CHAPTER 5

FRUITS OF THE NUCLEAR AGE

Questions surrounding energy—how much we need, where we get it, its impacts on the climate—are dominant factors in our lives, our politics, and world affairs. We measure energy in kilowatt hours, BTUs, or equivalent barrels of oil. Power, it is worth mentioning, is a measure of how much energy can be delivered in, say, a second, and is measured in units like watts and horsepower. In fact, James Watt, an early inventor of steam engines, invented the unit for which he's named and also introduced the term *horsepower*, as a marketing scheme for his engines. In any case, calling E the energy of an object, m its mass, and c the speed of light, $E=mc^2$ is probably the most famous equation there is. The speed of light is an enormous number (186,000 miles per second, or 300,000 kilometers per second). As a result, even a tiny amount of matter contains a vast amount of energy. A teaspoon of water contains enough energy to power a city for several days. But accessing this energy is another matter.

In an atomic bomb, the explosive is about 15 kilograms (33 pounds) of uranium; only a thousandth of that uranium is turned into energy, but the resulting explosion is horrifically destructive. In a nuclear reactor, one year of operation converts about 1 kilogram of matter into enough energy to power a large city. Our ability to do these things relies on the enormous forces between the particles in the nucleus of an atom. In our journey through powers of 10, zeroing in on the atomic nucleus takes us five orders of magnitude down from atoms, to things that are a tenth of a trillionth of a centimeter, or 10^{-13} cm, in size. The energy packed into these tiny spaces is enormous and is what enables the construction of atomic bombs and nuclear reactors.

For the better part of a century, this energy was also deeply mysterious, though now it is completely understood. The underlying theory, like quantum electrodynamics (QED), forms the second of the three pillars of the Standard Model and is known as quantum chromodynamics, or QCD. With QCD, we understand the forces and particles in nature on scales five orders of magnitude smaller than atoms, and the history of the universe starting at a millionth of a second after the big bang.

Nuclear Physics

Nuclear Physics—the very phrase makes many of us uneasy. It conjures visions of nuclear weapons, and nuclear power, with their much-debated benefits and obvious dangers. But the science of the atomic nucleus is a fascinating topic, which launched much of the physics of the twentieth century. In the previous chapter we encountered Marie Curie and Ernest Rutherford, who used nuclear radiation as a tool for the study of matter and contributed in important ways to quantum mechanics. They would also each play a critical role in the development of nuclear physics.

Rutherford's discovery of the nucleus drove the understanding of atoms, but it raised another big puzzle. If the nucleus is a collection of protons and neutrons, since the protons all have positive charge, they should repel each other. What might hold them together in such a tiny space? There needed to be some additional force, stronger than the electromagnetic force—much stronger. With, perhaps, a certain lack of imagination, this force was dubbed the strong nuclear force, or *strong force* for short. Understanding the nature of this nuclear force quickly became one of the great problems of physics. Over the next few decades, many features became clear. The force, most importantly, acts only over a short range. In other words, two protons are pulled together only if they are almost on top of each other. Separated even by a very small distance just a few times larger than the size of either proton, they feel only their electrical repulsion. But figuring out the force law—in the way the law is known for electricity and magnetism or gravity— proved hard.

Progress in this direction came from the work of the Japanese theorist Hideki Yukawa in 1934. Yukawa applied the then new ideas of quantum mechanics, and especially the uncertainty principle, to the question of the nuclear force. The uncertainty principle, in one of its forms, asserts that one can't know simultaneously the energy of a process and the time it takes, with arbitrary precision. For electricity and magnetism, the carrier of the force is the photon. The photon has no mass, so its energy can be (in light of Einstein's $E=mc^2$) arbitrarily small—as small as one likes. So the time the photon travels, say, from a proton and electron, can be as long as one likes. Since light travels very fast, this translates into the possibility that the forces of electricity and magnetism act over very large distances (gravity is similar). Yukawa reasoned that the short range of the nuclear force might arise if the force carrier was massive. Plugging in the numbers that appear in Heisenberg's uncertainty relation, he predicted a particle with mass about 1/8 the mass of the proton. This prediction, from the data then available, was not too precise, but there was a firm prediction that there should be a particle interacting strongly with atomic nuclei, and significantly lighter than the proton.

In the years before the Second World War, a candidate for Yukawa's particle was discovered in cosmic rays, also by Carl Anderson, who we encountered in the previous chapter. It had roughly the expected mass, but it didn't interact with nuclei in the expected way. Research in nuclear and particle physics was disrupted by the war, so only in the years afterward was it realized that this first particle was not Yukawa's meson. Called the "muon," this particle was very similar to the electron, but about 200 times heavier. After the war, Yukawa's meson was discovered, both in cosmic rays and accelerators. Actually, there are three such particles, one with positive charge, one negative charge, denoted by π^+ and π^-, and one electrically neutral, denoted by π^0. They have nearly the same mass. So there is both a triumph and a mystery here. The masses of these particles, known as pions, are about where the theorist Yukawa said they should be; but no theorist predicted that experimenters would discover a particle like the muon, and even if they had, they would not have guessed that its mass would be close to that of the pion.

Symmetry as a Guide to Nature

In mathematics, and in many of the sciences, there are two notions of *hard*. There's one we use loosely, referring to a hard problem on a homework assignment or exam, or a description in a book we find hard to understand. Here the obstacles can usually be overcome if we just keep going, get help from someone more expert than us, or do some training. But there is another notion of hard, which is especially easy to appreciate in our computer age. For these problems, one has a strategy to solve the problem, typically involving some mindless rules to do a computation. All that's required are lots of additions and multiplications, or other familiar operations. The clever, thoughtful part of solving the problem lies in coming up with the rules. A computer can be programmed to do the mindless part of the calculation. But it may take the computer a long time, because the problem is so complicated and there are so many operations to do. The time might be so long as to make the calculation impossible or the results useless. Forecasting weather can be in this class. If one demands too much precision of the calculations, your computer model may successfully predict a storm only after it happens. Such problems are hard in a technical or mathematical sense.

For reasons that will become apparent, the interactions involving the strong nuclear force are just such a hard problem. This was clear early on, even though it took some time to develop a strategy that would allow even an extremely powerful computer to solve the problem—and it would be decades before computer technology reached the required level. From the earliest days of the subject, more indirect strategies provided insight into the nuclear force. Particularly important was the discovery of *symmetries* of the strong interactions.

Sometimes, in art or architecture, we find symmetry attractive. Sometimes, a lack of symmetry or even a slight asymmetry is appealing. This may also be true of things we observe in nature—a face, a plant, a mountain. There are two kinds of symmetries we usually have in mind when we contemplate such objects. One is symmetry under rotation. If we rotate a jug about an axis through its center, it may look the same or nearly the same. Another very familiar one is symmetry under interchange of left and right. We may notice this symmetry—and ways in which it is not quite perfect—in the features of people close to us. This is also called symmetry under reflection, or often *parity*. These symmetries—and others—play an important role in our understanding of physical law.

Aesthetic appreciation of symmetry in nature surely dates back to ancient times. But beginning with Galileo and Newton, it was realized that the laws of nature themselves possess symmetries, and these have consequences for phenomena in the natural world.

Some of the symmetries seem so obvious at first as to hardly be worth remarking. Newton's laws of motion and his law of gravity don't carry any kind of time stamp. They are the same today as they were yesterday, or 1,000 years ago, and we'd be surprised if they were different tomorrow. This extends to all the laws we know, even when quantum mechanics is included— those of electricity and magnetism, those of the Standard Model. But the consequence of this fact is extraordinary—it is the *conservation of energy*. This law exerts a tyranny over our day-to-day lives. It tells us we can go only so far on a gallon of gas (or a kilowatt hour stored in the battery of an electric vehicle), that it takes a huge amount of energy to launch a payload into space, that we are tired after hard physical labor. We also don't need to know a lot to anticipate the consequences of this principle. We don't have to know the details of every chemical reaction in our car's engine, or the complicated mechanisms breaking down the food we eat to produce motion in our muscles, to understand and even quantify the outcomes.

For Newton's laws, this was known in the late eighteenth century, but the full generality of this connection between symmetry and conservation laws was not appreciated and codified until the early twentieth century, by the mathematician Emmy Noether. Born in Germany in 1882, Noether was an outstanding mathematician who would have been a prominent professor had she been male. Despite the limitations on her career as a woman mathematician, on her own and in collaboration with others she did outstanding work. Of her work on symmetries, Einstein wrote, "Yesterday I received from Miss Noether a very interesting paper on invariants. I'm impressed that such things can be understood in such a general way. The old guard at Gottingen should take some lessons from Miss Noether! She seems to know her stuff."

In 1933, Noether fled persecution as a Jew, taking a position at Bryn Mawr College in the United States. She died two years later. In a letter to *The New York Times*, Einstein wrote: "In the judgment of the most competent living mathematicians, Fraulein Noether was the most significant creative mathematical genius thus far produced since the higher education of women began. In the realm of algebra, in which the most gifted mathematicians have been busy for centuries, she discovered methods which have proved of enormous importance in the development of the present-day younger generation of mathematicians."

In her work, Noether made clear that other principles lead to conservation laws in addition to conservation of energy. For example, the laws of nature are the same whether I am standing in front of my office or in front of the office next door. More generally, they are the same if I move a bit to the left or right, forward or backward, or up and down. The consequence of these symmetries is the conservation of momentum, a principle as controlling as that of conservation of energy. Similarly, nature doesn't care if you are moving to the north, or north and slightly east. This is the symmetry of rotations, and it is a feature of all the laws of nature we know. It is reflected in a somewhat more subtle but equally profound principle, the conservation of angular momentum. This principle explains, for example, why the earth rotates around its axis at a steady rate over geological time scales.

Yet another symmetry explains the conservation of electric charge, one more sacred principle. This symmetry is far more subtle. It doesn't have to do with changes one can see or feel. But it's a built-in feature of Maxwell's equations, a feature that happily survives when we add quantum mechanics to the mix.

The exploration of symmetries has been a theme in particle physics almost from the beginning of the discipline. Careful checks were performed in the early accelerator experiments of the conservation of energy, momentum, and angular momentum. All have been shown to hold to extraordinary precision. More interesting, because it turns out *not* to be a perfect symmetry, is the symmetry under mirror reflection, also known as parity as mentioned above. We all have a good intuitive understanding of what parity is; indeed, we often have trouble distinguishing our left and right. We can be a bit more precise about what "parity invariance" would mean for the laws of nature. If we watch an event, and watch the same event as reflected in a mirror, both should make sense to us, i.e., both should appear to respect the laws of nature. This is a fact of Newton's laws and Einstein's general relativity, and a feature of the laws of electricity and magnetism, so it was for a long time taken for granted. But in the 1950s, as particle accelerators allowed detailed study of the radioactive decays of certain elementary particles, there were big surprises. It was hard to understand some of these decays if one insisted that parity was a good symmetry. People tried, hypothesizing the existence of pairs of particles nearly identical in every way. But Tsung Dao (T. D.) Lee and Chen-Ning (C. N.) Yang, two young physicists from China then working at the Institute for Advanced Study, made the more radical suggestion that perhaps parity is not a good symmetry of nature. They proposed that the study of certain radioactive decays of atomic nuclei could settle the question, and ingenious experiments to test parity were soon performed by Chien-Shiung Wu of Columbia University. Wu found, indeed, that parity is not a good symmetry. This violation of parity is an intrinsic feature of the Standard Model.

Lee and Yang received the Nobel Prize for their hypothesis of parity violation. On the other hand, "the Chinese Madame Curie," as she was known, who designed and performed the experiment that made the actual discovery, did not. While all three scientists were well known, and were a source of pride in the Chinese community in the United States and in China, Wu faced many challenges as a woman in science. Despite these challenges, Wu was eventually elected president of the American Physical Society, the principal professional society of physicists in the United States. Her accomplishments were recognized in 1975 when President Gerald Ford presented Wu with the National Medal of Science. She also emerged as a prominent advocate for human rights, in China and elsewhere.

Another important symmetry is *time reversal*. In Newton's world, make a video of some event—a ball rolling down a hill, a planet in its orbit around the sun—and then run the video backward. What you see may be surprising, perhaps challenging to arrange, but it obeys the laws of nature. Once we've given up parity as a symmetry, symmetries like time reversal would seem in jeopardy. In 1964, a small violation of time reversal symmetry was discovered in the weak interactions, of a very subtle form.

The symmetries we have so far encountered—translations in space and time, rotations, and parity—are symmetries we encounter in our day-to-day experience and of which we have some intuitive understanding. With the discovery of quantum mechanics, symmetries took on an even more central place in the laws of nature. Energy, momentum, and angular momentum were still conserved and their role, if anything, elevated. But quantum mechanics also revealed symmetries of a less familiar sort. Physicists call these symmetries, which have no obvious connection to space and time, "internal symmetries." Perhaps the first of these was suggested by Werner Heisenberg, one of the inventors of quantum mechanics, famous for his uncertainty principle. Heisenberg noted that the proton and the neutron have very nearly the same mass. In fact, their masses are equal to roughly a part in 1,000. This might be a coincidence, but it is pretty remarkable. On the other hand, the proton and neutron would seem dramatically different—one particle carries electric charge, one does not. Heisenberg reasoned that, if you turn off the electric charge, perhaps you couldn't tell these two particles apart—there is a *symmetry* that relates them. Since the number of neutrons in a nucleus determines which isotope of that type of an element one has, he called this "isotopic spin," or "isospin." He and others checked that the properties of nuclei were consistent with such a symmetry, again if one ignored the repulsive electric force between protons and the tiny difference in the proton and neutron masses.

Yang and Mills: Following Einstein, a New Type of Symmetry

C. N. Yang's role in the discovery of parity violation, the break-down of what to many, until that time, seemed a self-evident symmetry of nature, was remarkable. But slightly earlier, he made another discovery of profound importance. To appreciate this idea, we need to return to Einstein. One way to think about Einstein's general relativity is in terms of symmetries. We know the laws of nature remain the same if one rotates a system about some axis. But while this is a property of Newton's laws, it is actually strictly necessary to rotate *the whole universe* this way if the symmetry is to hold. This sounds a bit crazy, and it is. Einstein's general relativity liberates the laws from this requirement. Within Einstein's theory, one can rotate just a small part of the universe—in a lab, in my classroom, in the solar system. But this works only if one includes the gravitational field. The force of gravity, in other words, is forced upon us by this symmetry principle.

Yang and Robert Mills, in 1954, asked the same question about Heisenberg's isospin symmetry. Shouldn't one be able to trade a neutron for a proton anywhere in the universe, without having to do it everywhere? They wrote down a theory that implemented this idea. The mathematics was very pretty. Just as Einstein's insistence that one can implement a rotation locally in space-time implies the existence of the *gravitational field*, Yang and Mills's insistence that one can make an isotopic spin transformation anywhere in space-time leads to the prediction of three new fields, more similar to those of electricity and magnetism, and three massless particles, similar to the photon. No such particles exist, though Yang and Mills suggested that perhaps three massive particles known at that time might, somehow, be these particular "vector mesons." How these particles might grow mass was a mystery, and the theory, for more than a decade, remained dormant.

Yang went on to many profound accomplishments in the field of condensed matter physics and to general aspects of theoretical physics. For many years, he headed the C. N. Yang Institute for Physics at the State University of New York at Stony Brook. Despite his contributions to our understanding of particle physics and basic laws of nature, he eventually became a harsh critic of the particle physics enterprise. He is particularly unenthusiastic about "big science," of which particle physics is perhaps the epitome. As of this writing, Yang is a prominent opponent of proposals in China to build an extremely high energy accelerator (the "Chinese Collider"), a project requiring an investment comparable to the cost of the Large Hadron Collider at CERN in Switzerland, and of much higher energy.

The theory of Yang and Mills, as we will see, has emerged as the underpinning of the Standard Model. It has also had profound impact in mathematics. But making sense of these theories and seeing their possible role would take over a decade. Much happened in those intervening years.

Why Are the Pions Light?—the Nambu-Goldstone Phenomenon

With the discovery of the pion, Yukawa's picture appeared to give at least a rough description of the strong nuclear force. But it was not satisfying in the way the quantum theory of electric and magnetic forces (QED) was. First, precisely because the interactions were so strong, one couldn't apply the methods of Feynman, Schwinger, and Tomonaga to this problem. So it was hard to establish what the theory predicted. But it also became clear rather quickly that the theory could not be complete as a model of the strong interactions. The discovery of the pions was followed by the discovery of other, heavier particles, which also seemed to play some role in the nuclear force. As a group, the strongly interacting particles were called *hadrons*. The questions now became: What is the role of all these other particles? And why are the pions somehow special? In particular, why are they significantly lighter than any of the others and the most important in explaining the force between neutrons and protons?

A solution came in the work of the great Japanese-American physicist Yoichiro Nambu. He had come to the United States as a young man, taking on a position at the Institute for Advanced Study in Princeton. While always a modest and gentle person, Nambu was not shy or retiring. When he arrived at the Institute, J. Robert Oppenheimer, who was then the director, told the new members (the title of postdocs at the Institute) that they were not to disturb the great Einstein. Nambu acknowledged the instruction and then immediately made an appointment to meet with Einstein. He was eager to acquire whatever wisdom the legendary physicist would be willing to impart. At their meeting, Einstein complained that none of the younger people came to see him. Nambu, somewhat unusually for that time, and particularly for a young person coming from abroad, owned a car, and after that first meeting, he made a point of offering Einstein rides to work. He surreptitiously took a picture of Einstein one morning walking toward the car, not a small feat with the cameras of those days. In our day, this would have gone viral on social media.

In any case, after his time at the Institute, Nambu went on to the University of Chicago. There he did a range of important work, but he was particularly interested in the symmetries of the strong interactions and their connection to the pions. Nambu considered the possibility that the strong interactions had a symmetry of a then unfamiliar type. Unfamiliar in two ways. First, because it emerged from the combination of relativity and quantum mechanics. If the electron had no mass, he realized, it would obey a curious conservation law. The spin of the electron points in some direction. In quantum mechanics, if the electron were massless, it would, like the photon, move at the speed of light. The direction of its spin might be along the direction in which the electron of its motion, or in the opposite direction. If along the direction of motion, it will stay that way; similarly, if opposite. This is the conserved property. The symmetry connected with this conservation law is called "chiral symmetry" because it has to do with handedness (the word *chiral* comes from the Greek word for "hand"). The electron is not massless (though at very high energies, its mass can sometimes be ignored and the conservation law holds), and the proton and neutron are certainly not.

Here came the second, great leap in Nambu's thinking. He conjectured that the strong interactions have such a symmetry, but it is a *broken symmetry*. At first, this idea may seem a bit weird, but the notion of a broken symmetry is actually quite familiar. The laws of nature are symmetric under rotations, but objects we encounter are typically not. The handle on a water jug, for example, breaks the rotational symmetry. This is not paradoxical; the underlying symmetry is reflected in the fact that we can orient the jug so that its handle lies in any direction. Objects other than perfect spheres have some intrinsic direction and, if sitting at rest, must point in *some* direction. Physicists say that such a symmetry is "spontaneously broken." Nambu reasoned that chiral symmetry in the nuclear forces is of this type. He also realized that if a symmetry of the laws of nature is broken in this way, there is a consequence: There must be a massless particle. There are no massless particles in the strong interactions, but the pions are much lighter than the others, and he identified these as the candidate massless particles. If the chiral symmetry is not an exact symmetry of the underlying laws of the strong interactions but is broken "a little bit," this would account for the small masses of the pions. Jeffrey Goldstone of MIT proved a general result that massless particles arise from spontaneous symmetry breaking, so the massless particles are known as Nambu-Goldstone bosons. Extensive experimental study verified that the pions behave in just the way expected of such objects.

Lots and Lots of Strongly Interacting Particles

The late 1940s saw the beginning of the accelerator era in particle physics. While primitive accelerators had been built in the years before World War II, physics was now viewed as an important part of the nation's defense, and generous funding became available from the federal government. At the same time, technological developments during the war, and the training of a cadre of skilled people, both scientists and technicians, facilitated the boom. Accelerators were built and operated at UC Berkeley, the Brookhaven National Laboratory on Long Island, and elsewhere. The energies of these accelerators were more than enough to produce pions and soon yielded a raft of other particles as well. The details, for our purposes, are not important. What is important is that there were hundreds of these particles. Like the pions, these new particles were all short-lived, decaying in many cases in about 10^{-20} seconds! But they were sufficiently distinctive that their properties could be measured with precision.

Yet, now there was a mystery. Initially the proton, the neutron, and the pions were thought to be fundamental, structureless objects, like the electron. But the proliferation of newly discovered particles called this picture into question. Careful measurements, in fact, showed that the proton has a size and shape similar to an atomic nucleus, about 10^{-13} cm. So perhaps, just as the elements of the periodic table are built of electrons and nuclei, these new particles were built of some other entities. The way out of this confusion was provided by Murray

Gell-Mann, who established an analog of the periodic table for the hadrons. His table was based on what was then, for most theoretical physicists, an unfamiliar branch of mathematics called *group theory*. The number eight played a crucial role in the table. The lightest mesons came in eight types, as did the lightest baryons—strongly interacting particles of spin 1/2 like the proton and neutron. Gell-Mann (who passed away in 2019) was quite erudite—and liked to let people know it—so showing off his interest in languages and Eastern religions, he called this the Eightfold Way. In Buddhism, the Eightfold Path is the path to nirvana, comprising eight aspects in which an aspirant must become practiced: right views, intention, speech, action, livelihood, effort, mindfulness, and concentration. In Gell-Mann's periodic table, the eight referred to various more prosaic properties, such as electric charges of the particles.

Dmitri Mendeleev had put forward his periodic table of the elements based on regularities in chemical properties. Only with the discovery of quantum mechanics were its features understood in terms of properties of electrons and atomic nuclei. Gell-Mann, simultaneously with George Zweig, implemented this second step for the hadrons: They proposed that the Eightfold Way could be understood in terms of particles analogous to the electron, protons, and neutron in atoms, called *quarks*. Gell-Mann chose the name from a passage in James Joyce's *Finnegan's Wake* (Zweig called these objects Aces, but the name never stuck). Initially, there were three types of quarks, rather playfully called up, down, and strange.

The quarks worked beautifully to account for the properties of the strongly interacting particles—the hadrons. But they had some peculiar properties. They carried electric charge. From a human perspective, the electron is the most important of charged particles. It is what makes up electric current and is used to store information in computers or our cell phones. Because it is so important, it is natural to take the charge of the electron as a basic unit. By a convention that traces back to Benjamin Franklin, we will say that the electron has charge minus one. The proton, then, has charge plus one, so that atoms are electrically neutral. The quarks hypothesized by Gell-Mann and Zweig would have charges that were fractions of this basic unit: 2/3 or minus 1/3. Particles like the pion, with spin zero, in the quark model consist of a quark bound to an antiquark. These particles are known as mesons. Particles like the proton and its excitations, which carry spin like that of the electron, are bound states of three quarks. The proton, with charge one, was composed of two up quarks and one down quark. The neutron, with charge zero, was composed of one up and two down quarks. Other hadrons were built of other combinations. The π^+ meson, for example, consisted of an up quark and an *anti*-down quark (the antiparticle of the down quark).*

While this new periodic table worked well, there was a puzzle. Electrons are easily ejected from atoms. We're used to this from seeing sparks when a strong electric field (in a circuit, for example) rips electrons from atoms. Chemists work all the time

---

*Of the first three quarks, the up quark had charge two-thirds, the down and strange quarks had charge minus one-third. The proton was built of two up quarks and a down quark, giving it charge one; the neutron, of an up quark and two down quarks, adding up to charge zero. The mesons, particles of spin zero or one, were in some cases charge one (a combination of a quark and an anti-down quark) and in some cases charge minus one or zero.

with ions, and scientists study electrons and nuclei individually in many controlled situations. But for quarks, nothing similar happens. When protons collide with each other, with neutrons and with pions, no objects with fractional charge—charges plus or minus 1/3 or 2/3 that of the electron or proton—are observed in the debris. Scientists searched for such fractional charges all over the place, even in moon rocks. Gell-Mann himself retreated, for a time, to the view that quarks were just some sort of useful mathematical plaything, with no physical reality.

But a different kind of experiment eventually established that quarks *are* real. In the mid-1960s, Richard Feynman at the California Institute of Technology (Caltech) and James Bjorken (known as BJ) at the then newly built Stanford Linear Accelerator Center (SLAC) started to think about what might happen in very high energy experiments if protons and neutrons consisted of quarks. Feynman, whose competition with Gell-Mann was notorious, refused, at least for some time, to call the constituents of the hadrons quarks, instead calling them partons. In any case, Feynman and Bjorken made a sharp prediction. Experiments similar to Rutherford's, where electrons scattered off of nuclei, at very high energies should reveal the internal structure of the proton and neutron. A sequence of experiments at SLAC uncovered just this phenomenon. The proton was seen to be built of particles with fractional charges. This work won the Nobel Prize for Jerome Friedman, Henry Kendall, and Richard Taylor. For many, Henry Kendall is a familiar name; he went on to found the Union of Concerned Scientists, which has done notable work on issues of nuclear power and on environmental and energy policy more generally.

So it would seem that physicists were poised to uncover new laws of nature, those which governed the nuclear force. But the very successes of the quark model posed serious challenges. One might have thought that relativity and quantum mechanics required that the nuclear forces should be described by a quantum field theory. But it seemed that no quantum field theory had the required properties—either to explain why quarks were not seen free, isolated by themselves, or why hadrons would behave as if they were composed of quarks when banged together hard enough.

The Breakthrough: Yang-Mills Theories and Their Remarkable Properties

The explanation of the first puzzle, that the hadrons look like collections of quarks when collided hard together, was provided by David Gross and Frank Wilczek at Princeton and David Politzer at Harvard in 1973. They all realized that success in explaining this first point required that the theory have a property called *asymptotic freedom*. This is a fancy and rather colorful way of saying that the force between quarks must grow weaker as the quarks come closer together. But this didn't seem to happen in the quantum field theories then familiar to theorists. In quantum electrodynamics, for example, one finds the opposite behavior: The force gets stronger as electrons approach each other. There was even an argument, seemingly based on general principles of quantum mechanics, that this would always be the case.

Gross and his then graduate student Wilczek, in fact, set out to prove that no known quantum field theory had this property. Politzer pursued the same problem but with a different outlook, at the suggestion of his thesis adviser, the late Sidney Coleman (who we'll meet again later). For theories like QED and Yukawa's meson theory, the calculations were relatively easy and familiar. But for one class of theories the problem was more challenging. These are the theories of Yang and Mills, known today as *non-Abelian gauge theories* or *Yang-Mills theories*, which we encountered earlier. For a decade, these theories had languished. They were fascinating but hard to understand, and no one put forth a convincing role for them in understanding the laws of nature. While there had been some progress, they were not understood at the level at which QED was understood. Richard Feynman, much as he had in the late 1940s for QED, guessed a set of rules for performing calculations in these theories. The role of Freeman Dyson in making sense of Feynman's guess was played, at first, by two Soviet physicists, Ludvig Faddeev and Victor Popov. In a real sense they outdid Feynman. In figuring out the quantum mechanics of Yang-Mills theories, they took one of Feynman's more outlandish early ideas and turned it into a useful and powerful tool.

But still the calculations were hard. The resulting mathematical expressions didn't make much sense. A further breakthrough came with the work of two Dutch physicists, Martinus "Tini" Veltman and his student Gerard 't Hooft. An expert on calculating things in QED, Veltman was perhaps the first to develop computer codes to calculate in the theory, not just by adding up long columns of numbers but by doing algebra, as humans do. Algebra can be hard, and in the QED problem, there can be *lots* of it. It is also rather mindless, a perfect task for computers. But it was his student, 't Hooft, who drove much of the needed development for Yang-Mills theory. For many years, in fact, tension existed between the two of them over who should receive credit for the most innovative aspects of this work. They shared, however, the Nobel Prize for their elucidation of the quantum mechanics of Yang-Mills theories in 1999.

Thanks to the work of Faddeev and Popov and 't Hooft and Veltman, Gross and Wilczek and Politzer could perform the required calculations. As so often happens, at this early stage, these calculations were quite challenging; now they are a standard homework assignment in my graduate physics course. They discovered that these Yang-Mills theories have the required property.

With their discovery, the quark model, combined with Yang-Mills theory, led naturally to a real theory of the strong force, a new set of laws of nature, built on the ideas of Yang and Mills. Hadrons consist of quarks coming with an intrinsic property called color. There are three of these colors, sometimes taken to be red, blue, and green. These particles interact through the exchange of eight types of particles similar in some ways to photons, called gluons.

This new theory was called, by analogy to quantum electrodynamics (QED), quantum chromodynamics (QCD). Not only did Gross, Wilczek, and Politzer explain the rough features of the SLAC results, but they also predicted small corrections to the predictions of Bjorken and Feynman. With this began a long process of testing the theory, and also extending the class of processes for which the theory *could* make predictions. Many challenges lay ahead to fully understand and test this theory, but by now, this pillar of the Standard Model has been verified in great detail.

The work of Gross, Politzer, and Wilczek was awarded the Nobel Prize in 2004. Some of us feel that Sidney Coleman was also deserving of this recognition, but at most three people can share the prize.

With the problem of asymptotic freedom solved, there was now the question of quark confinement to understand. The property of asymptotic freedom offered some hope. The flip side of the fact that the strong interactions become weaker when quarks are close together is that they become stronger as they move far apart. Perhaps, it was thought, they become so strong that one simply can't pull quarks apart. This property was dubbed "infrared slavery." But the problem of quark confinement has proven to be a hard one, in the technical, mathematical sense. Indeed, this was, in many ways, a harder problem to tackle than theoretical particle physicists had previously encountered. They looked to other fields where analogous phenomena arise. The late Ken Wilson of Cornell first turned the abstract problem into a precise, if extremely hard, question. In a way similar to Faddeev and Popov, Wilson took the same crazy idea of Feynman's for doing quantum mechanics and did something crazier with it, which suddenly turned it into an object so sensible that one could put it on a computer. Wilson proposed that, in order to obtain manageable computations, one replace the space-time continuum with a set of discrete points. Mathematicians and physicists call such a space-time a *lattice*, thinking of the sort of regular structure one has in decorative arts and elsewhere. A picture of such a lattice appears below. With this, the problem of solving the color gauge theory—QCD—became a well-posed computer problem. This problem, however, is *hard*.

To give some sense of what hard means, if the lattice has 100 points, one has to do something like ten to the ten to the eighth power operations ($10^{10^{8}}$) to obtain an accurate result. If we built a computer using every electron in the universe, we couldn't do these computations. So a great deal of cleverness has been required to reduce the problem to a manageable scale. Even then, one must use extremely powerful and expensive computer networks. Researchers built dedicated machines just for this problem. But eventually the growth of computer power and the cleverness of the algorithms to do the computations converged, and by the beginning of the new millennium, reliable results were available. Not only did these computations exhibit quark confinement, but details of the hadrons were reproduced correctly—quantities like the masses of the proton and neutron and the pions (the pions turned out to be particularly hard). Even better, it became possible to calculate quantities for experiments not yet done, especially involving the heavy b quark.

Many of the characters in this story made other major contributions to theoretical physics. While as of this writing, C. N.

A two-dimensional lattice

Yang is ninety-eight years old (born in 1922), he is a force in Chinese science. Gerard 't Hooft continued to play a leading role in the understanding of the strong interactions. Along with Leonard Susskind and Stephen Hawking, he has been among the most profound thinkers on the problem of developing a quantum theory of gravity. David Gross later played a crucial role in the development of superstring theory. As a result of his work on computation in the strong interactions, Ken Wilson developed a deep interest in the problems of large-scale computing generally, studying applications in chemistry and elsewhere. He won the Nobel Prize for his work both in condensed matter physics and in the strong interactions in 1982.

But for many physicists—theorists and experimentalists—there is something not entirely satisfactory about all this. Physicists like to make back-of-the-envelope calculations, in other words to do something simple, to have at least a rough understanding of why detailed and complicated calculations work out the way they do. They are not totally satisfied with the statement, "Well, the computer did this really hard computation and here is what came out." The Clay Mathematics Institute of Peterborough, New Hampshire, sponsors a set of prizes for hard problems in mathematics and physics known as the Millennium Prizes. One of them is for a demonstration that quark confinement occurs in QCD which doesn't rely on computers. I tell my students about this one. The prize carries a cash award of $1 million. It is still unclaimed.

CHAPTER 6

THE WEIGHT OF
THE SMALLEST THINGS

One of the key qualities scientists use to characterize an object is its mass. Gold enjoys its special status partly because it is rare and shiny, but also because even small amounts are heavy. By contrast, the masses of the elementary particles are extremely tiny. A thousand trillion trillion electrons would weigh a little over a gram; that number of protons would weigh about a kilogram. Not only are we now able to weigh large collections of these particles, we can study them one at a time and measure their masses individually. Even the heaviest we know, the top quark, is so light that a trillion trillion of them would weigh about a kilogram. But we couldn't put a collection together to weigh them. The top quark has a half-life of about 10^{-24} seconds. The collection would almost immediately disintegrate into other, lighter particles. We have to measure the mass of the top quark more indirectly. The same is true of measuring the lifetime of the top quark—we don't have clocks that measure such tiny time intervals. While not quite as dramatic, there is a similar story for the masses of many of the other particles.

What accounts for the masses of these particles? Are these just numbers in the table at the back of some textbook or that we look up on the internet? Or can we understand them in terms of some set of fundamental principles? We've already seen two examples of such principles at work: Dirac explained why the positron, the antiparticle of the electron, exists and has *exactly the same mass* as the electron. The pioneers of QED showed that this was the same for antiparticles of every particle. The photon has no mass as a consequence of an underlying principle of electrodynamics. But what about all the other particles? To address this question, we need to put together another important piece of the Standard Model.

Radioactivity is frightening—in large quantities, such as near an exploding nuclear weapon or an out-of-control nuclear reactor, it can kill on the spot; smaller exposures can cause cancer. In some sense, though, radioactivity is actually less frightening than other poisons. From the experience of Hiroshima, Nagasaki, Chernobyl, and Fukishima, we know quite reliably the effects of radiation in large doses; this is not always the case for chemical and environmental poisons.

Radioactivity occurs naturally in materials on the earth, or when cosmic rays, the high-energy particles produced in distant regions in the cosmos, strike the upper atmosphere. The Curies and Rutherford used different forms of naturally occurring radioactivity to probe the structure of matter. Rutherford classified three types of naturally occurring radioactivity:

alpha, beta, and gamma radiation. Alpha rays he eventually understood to be helium nuclei, emitted when certain heavy nuclei, including uranium and radium, fall apart. Gamma rays are high-energy photons—much more energetic than those associated with light, radio waves, or the photons in your microwave oven. Beta rays consist of electrons or their antiparticles.

Understood this way, the harms that radiation causes are not so mysterious. Each of these types of particles collides with atoms and molecules in our bodies, causing them to break apart. In intense doses, this can damage large quantities of tissue and kill quickly. In smaller doses, the breakup of DNA, for example, can set the stage for a range of illness. Understanding doesn't make these consequences less frightening but can at least alert us to potential harms and strategies to avoid or mitigate them.

Prior to the discovery of the neutron, it was thought that a typical nucleus contained protons and some number of electrons, canceling out some of the proton charge. But it was not clear what would hold the electron so tightly to the proton— why they would be closer together than in hydrogen. With the neutron's discovery, not only was the identity of a large fraction of matter understood, but the phenomenon underlying beta radiation became clear. A neutron, in isolation, is radioactive. It falls apart, typically in about eleven minutes, to a proton, an electron, and something else—a neutrino, the very light particle we have already encountered, which hardly interacts with other forms of matter. Recall that the neutron is ever so slightly heavier than the proton. The difference in their masses is only slightly larger than the mass of the electron. According to

Einstein, mass and energy are equivalent. So since energy is conserved, there is, as a result of this close similarity of the proton and neutron, just barely enough energy for this decay to occur. In a nucleus, the neutron has even less energy, and for many nuclei, the decay does not happen at all. Carbon 12, with six protons and six neutrons, (fortunately) is stable, permitting life. Carbon 13, with six protons and seven neutrons, is unstable.

The process can also go in reverse. An electron can strike a proton, the pair turning into a neutron and shedding a neutrino. In this way, one type of nucleus can turn into another. This is crucial to nature's alchemy. It is how stars, made mostly of hydrogen, can produce nuclei of heavier elements— eventually carbon, oxygen, iron, and other elements crucial to life. These nuclei are spread around the universe by stars that undergo explosive death—supernova explosions. We are made of the products of the dust of early generations of stars.

Returning to neutron decay: Eleven minutes may not seem a very long time, but, in a sense, it is long—ridiculously long. Humans created units of length like inches, feet, miles, kilometers from distances we experience in our day-to-day lives. But there are other, more natural, ways to think about these various lengths and times. The time it takes light to reach the nearest star to our sun is about 2.4 years. The time it takes light to reach the moon is about 2 seconds—about seven powers of 10 less. The time it takes light to cross an atom is 10^{-18} seconds. The time it takes light to cross a neutron is about 10^{-23} seconds! Compared to that, the neutron lives essentially forever. This very long time means that the force associated with beta decay are extremely feeble. This force is called the weak force, and interactions involving this force are called weak interactions.

New forces suggest new laws. In 1933, starting with what was then known of neutron decay, Enrico Fermi proposed a new set of laws: those which described the weak interactions. Fermi took his cues from QED. The theory reproduced many features of neutron decay, but it was not complete. It was a long time before Fermi's theory could describe all the beta decays of complicated nuclei and unstable hadrons. Understanding these decays in terms of quarks, electrons, and their relatives, as well as neutrinos, took the better part of four decades.

But even as these features fell into place and the theory came to describe a wealth of data, there was a huge problem with Fermi's theory. If one studied the theory at ever shorter distance scales or higher energies, it ceased to make sense. As a quantum mechanical theory, it predicts probabilities. When the processes involved distances about 1,000 times smaller than the size of a proton, many of these probabilities were greater than one—what could that possibly mean? A fix for this was clear early on—already, in fact, to Fermi. In QED, the force between charged particles arises from the exchange of photons. In Einstein's theory, the gravitational force arises from the exchange of gravitons. Both photons and gravitons have no mass. As a result, they can easily travel very long distances, and the forces can act over great distances. The weak force, in Fermi's theory, acts only over a very short distance—infinitely short, in fact. This statement—not quite as crazy as it may sound but still somewhat crazy—is the source of the problems. If the weak force, like electromagnetism and gravity, was mediated by the exchange of particles, but these particles were massive, then they would travel only short distances, accounting for the very short range of the force. This would be similar to the role of pions in the nuclear force. To reproduce the successful parts of Fermi's theory, these particles would have masses 100 or so times the mass of the proton or neutron, and two of these particles would be needed—heavy versions of the photon carrying electric charge. These particles were known as the W bosons, W^+ with charge opposite to that of the electron, W^- with the same charge.

By themselves, though, these particles did not fix all the problems of Fermi's theory. The breakdown of the theory was postponed to still shorter distances. But the theory broke down nonetheless. There were, first, problems with these massive versions of the photon. We've said that the photon itself is massless, due to the underlying symmetry principles of QED. It turns out that QED holds together as it does only because of these symmetry principles, so-called gauge symmetries. The situation is even more acute in the theories of Yang and Mills, where such charged force carriers ("vector bosons" or "gauge bosons") would seem fine, but only if they were massless as well, due to an even larger set of gauge symmetries.

All this changed in 1964, when several theorists discovered that there is a realization of the gauge principle where the gauge bosons are massive. Those who worked on this problem included Peter Higgs in Scotland, Robert Brout and Francois Englert in Belgium, Gerald Guralnik and C. R. Hagen in the United States, and Tom Kibble in England. As a result of historical accidents, this manifestation of the gauge symmetry came to be known as the *Higgs phenomenon*. The Higgs discoverers—I'll refer to them as the "gang of six"—found that if broken symmetries like those of Nambu and Goldstone are gauge symmetries, there are no massless bosons. Instead, the gauge bosons themselves are massive.

There were two elements to the theory written down by the gang of six. First, there was a new type of field, the *Higgs field*, of a sort different from those of electricity and magnetism. The Higgs field is a scalar field, which refers to the fact that, unlike the electric and magnetic fields, which point in a particular direction (the magnetic field of the earth points toward the north pole), the Higgs field has no directionality. What is important is that this field takes a nonzero value everywhere in space, thus giving rise to the masses of the elementary particles; the larger the Higgs field, the larger the masses. Second, just as the electromagnetic field is associated with a particle, the photon, and the Yang-Mills fields are associated with gluons and the W and Z bosons, the Higgs field is associated with a particle, in this case a particle without spin, known as the Higgs particle.

Nambu, who did so much for the development of particle theory, played a crucial role in this story as well, though distinctly under the radar. In the original version of the paper that Higgs submitted for publication, while he understood that his mechanism gave the gauge bosons a mass, he did not realize that his model predicted the extra scalar—the *Higgs boson*. It was the referee of the paper—Nambu—who pointed this out.

The Higgs discoverers did not actually build a theory of

weak interactions using their observations. This was achieved in 1967 by Sheldon Glashow of Harvard, Steven Weinberg (then at MIT, later at Harvard and the University of Texas), and Abdus Salam, a distinguished Pakistani physicist then working at Imperial College, London. Their theory became the part of the Standard Model that describes the weak and electromagnetic interactions. They made a particular proposal for the implementation of the Higgs mechanism. This was the simplest of all possibilities. Their version of the theory contained two major additions beyond Fermi's version. In addition to the particles W^+ and W^-, there was another gauge boson, without electric charge, Z^0, and a scalar particle—essentially the one pointed out by Nambu. Their theory did not predict exactly how heavy these particles should be, but one could at least make a rough guess.

The various elementary particles that don't experience the strong interaction are known as the leptons. At that time, leptons included the electron, the muon, and the two known neutrinos. They fit nicely into the new theory. Weinberg's paper, in fact, carried the title "A Model of Leptons." The three quarks that were then known (up, down, and strange) did not fit so nicely. Glashow and two collaborators proposed that this problem could be solved if there were a fourth quark, which he dubbed the *charmed quark*. With these ingredients, the Higgs particle accounted not just for the masses of the W and Z^0 particles, but for all the quarks and the leptons.

While the discovery of the Higgs mechanism by the gang of six was impressive, Glashow, Weinberg, and Salam were already well-established theorists and would play a dominant role in theoretical physics for many years. Glashow and Weinberg, friends and rivals, were classmates at the Bronx High School of Science in New York City, graduating in 1953. In my days as a professor at City College in Harlem, I had two colleagues who were in that same high school class. One, Myriam Sarachik, clearly found her two classmates intimidating. A condensed matter experimentalist, she struggled as a female physicist in the early 1960s, coping with overt sexism in an era when such behavior was viewed as normal. She has many accomplishments to her credit and was already in those days honored with membership in the National Academy of Sciences. In recent years she has served as the president of the American Physical Society, the main US professional society of physicists, and in 2019 was the fourth recipient of the society's Medal for Exceptional Achievement in Research.

I first met Sheldon Glashow when I was a college senior and visited Harvard, hoping to do graduate study there. I would sympathize with what I heard from Myriam in later years—my brief encounter with Glashow, in which he asked me why I would imagine I was competent to do theoretical physics, only left me feeling like an idiot. I was quite fortunate that Harvard rejected my application. Three years later, Tom Appelquist, himself a refugee from the cutthroat environment of Harvard in those days, would become my thesis adviser. He was a wonderful teacher, and much more tolerant of my constant bumbling and confusion than those at Harvard would have been.

My encounters with Steven Weinberg, later in my career, were much more rewarding. Weinberg also didn't suffer from a low opinion of himself, but he was much more tolerant of lesser mortals than Glashow. He approached younger and less renowned colleagues as people worth talking to, people he could teach and learn from. By the time he made his great leap in understanding of the weak interactions, he already had a long list of accomplishments. He would remain a leader in the field for another three decades. He eventually left Harvard for the University of Texas at Austin, along with his spouse, Louise Weinberg, a professor of law. Weinberg passed away as this book was nearing completion. He will be greatly missed.

Abdus Salam also went on to other significant accomplishments. Educated in Pakistan and at Cambridge University, he played a leading role in Pakistani physics until 1964. A member of the Ahmadiyya Muslim sect, he left the country in 1974 when the sect was declared non-Muslim. Interestingly, there is an active Ahmadiyya community where I live in San Jose, and my wife and I have celebrated iftar with them on several occasions. Relations with the larger Muslim community in our area, at least, seem respectful.

Returning to our story, the model of Glashow, Salam, and Weinberg did not immediately take the community of physicists by storm. The problem was that the model was based on the ideas of Yang and Mills, which, as we have seen in our discussion of the theory of strong interactions, did not at first make sense in an obvious way. With the addition of the Higgs mechanism, things were even less clear. But Gerard 't Hooft solved this problem as well. Extending the methods he developed for the Yang-Mills theory to include the Higgs, he showed that the whole theory made sense. It seems that only at this point did Weinberg embrace his own theory.

We have seen that the theory predicted five new particles: the W^+, W^-, Z^0, the charmed quark, and the Higgs boson. Indirect evidence for the Z^0 was found at CERN and at the Fermi National Accelerator Laboratory (Fermilab) in Batavia, Illinois. The charmed quark was discovered in November 1974, by large research teams led by Burt Richter at SLAC and Sam Ting, of MIT, at the Brookhaven National Laboratory on Long Island. The dramatic signal they found launched what came to be called the "November revolution." It had been anticipated by Tom Appelquist, my adviser (even so, he wasn't granted tenure at Harvard). The jockeying for the Nobel Prize started shortly afterward. Glashow and Weinberg, in particular, were sure of their model and made a point of disparaging all the many competing models. As best I can tell, they called the theory by the rather bland-sounding name Standard Model because of their insistence that this should be *the* standard. By the late 1970s, in fact, a host of predictions of the model had been confirmed. The W^+, W^-, and Z^0 particles were themselves observed as particles in 1983. The experimental discovery of the charmed quark yielded a Nobel for Richter and Ting, as did the discovery of the Ws and Z for Carlo Rubbia and Simon van der Meer of CERN, for driving the development of the accelerator and detectors that led to their discovery.

Dramatic developments continued for the next thirty years. Even before the Ws and Z were seen, another set of quarks and leptons, almost unanticipated, were found. A partner of the electron and the muon, known as the tau (τ) lepton or sometimes the heavy lepton, was discovered at SLAC. The tau lepton is indeed rather heavy, almost twice as massive as the proton, and

3,500 times as heavy as the electron. The word *lepton* comes from the Greek, meaning "small," so calling the tau the heavy lepton is a bit of an oxymoron. Another quark, called the b quark (for *beauty* or *bottom*) was also discovered about this time. The symmetries of the Standard Model required one more quark, called the top. Another two decades of pursuit were required, but the particle was found in experiments at Fermilab in 1995. The top was far heavier than any of the other quarks or leptons—more than 10,000 times heavier than the electron, and 40 times heavier than the next heaviest quark, the b quark.

This third set of quarks and leptons was not really a total surprise. It had been anticipated by two Japanese theoretical physicists. Again, the prediction of this third generation followed from considerations of symmetry. We have mentioned the symmetry of time reversal of Newton's laws—the past and the future are identical, as far as the laws of nature are concerned. Newton's laws also don't know the difference between left and right—they are said to be parity symmetric, and both these symmetries are also features of Maxwell's equations. Once parity violation was discovered, the symmetry of time reversal was also open to challenge. For elementary particles, due to general principles of quantum mechanics and relativity, time reversal relates particles to antiparticles, and the symmetry is often called CP. In QED, both time reversal and parity turn out to be automatic, consequences of the other requirements on the theory, but it is not necessarily a feature of the weak interactions (or, we'll see later, the strong). At a point when the weak interactions were not well understood, the experimentalists Jim Cronin and Val Fitch studied this question.

The experiments involved a beautiful quantum mechanical phenomenon involving the K mesons, which we encountered in the previous chapter, and exploited a variant of the phenomena that so concerned Einstein, Podolsky, and Rosen. Cronin and Fitch established that time reversal is not a good symmetry of the weak interactions. It turns out that in the Standard Model as put forward by Glashow, Weinberg, and Salam, with two generations of quarks and leptons as in QED, time reversal is automatically a working symmetry. In 1973, Makoto Kobayashi and Toshihide Maskawa realized that with *three* generations or sets of quarks and leptons, this was no longer true. This seemed a minimal way to understand time reversal violation. Also, as we'll see, one can't understand one of the most basic facts of the universe if time reversal is an exact symmetry: The universe consists of matter, not equal amounts of matter and antimatter.

There were other possible explanations of this breakdown of symmetry; for a while, I was an advocate of an alternative. But at the start of the millennium, a series of beautiful experiments in the United States and Japan firmly established that the model of Kobayashi and Maskawa was correct. They were awarded the Nobel Prize for this work in 2004.

The Hunt for the Higgs Boson

So, by 2004, all the features of the Standard Model had fallen into place but one: the Higgs particle. Now the efforts of theorists and experimentalists turned to this missing piece. In 1990, my Santa Cruz colleague Howard Haber, along with Sally Dawson (Brookhaven National Laboratory), John Gunion (UC Davis), and Gordon Kane (University of Michigan), wrote a book, *The Higgs Hunter's Guide* (for those not old enough to remember, the title was a play on the name of the science fiction series the Hitchhiker's Guide to the Galaxy). Their book set out strategies for searching for the Higgs. It turned out that depending on the mass of the Higgs, the mechanisms by which it would be produced in accelerators and the ways in which it would decay were quite different.

Still, at a sequence of particle accelerators over three decades, there was no sign of the Higgs. By the time the Large Hadron Collider (LHC) began operation in 2008, all one could say for certain was that the Higgs was larger than about 116 times heavier than the proton.

The LHC was conceived in the late 1980s, at a time when the United States was pursuing the largest accelerator project yet undertaken, the Superconducting Super Collider (SSC). The SSC was to be built in Texas, near the town of Waxahachie, thirty miles from Dallas. The accelerator was to be a huge ring, fifty-four miles in circumference. The ring would consist of two tubes, each carrying a beam of protons circulating in opposite directions. To keep the protons in their circular orbits would require large magnets.

Magnets are familiar to us from toys and our refrigerators. Most of these are called permanent magnets. They are based on the fact that the spin of an electron gives rise to a small magnetic field, aligned with the spin. Most of the time, these spins point in random directions, and the magnetic fields of the many atoms in an object cancel out. Iron, and a few other materials, are special, in that the spins tend to line up (this is another example of spontaneous symmetry breaking). But magnetic fields are actually ubiquitous in nature and more commonly arise from electric currents. These sorts of magnetic fields have been critical to particle accelerators from the beginning. What's important for accelerators is that the paths of charged particles, passing through a magnetic field, bend. Magnetic fields can be used to steer particles through the accelerator, to make them strike a target, and also to measure their speeds.

The strongest magnets scientists and engineers know how to make are built of superconducting materials. In your house, your car, your workplace, electricity is carried by materials that carry electricity readily. But you may notice that the electrical devices you use often become hot after a little while. This is because of *resistance*. Most materials resist electrical flow; copper and aluminum are relatively good conductors, which means they don't resist the flow so much, and they are not terribly expensive. In 1911, the Dutch physicist Heike Kamerlingh Onnes discovered that some materials, when cooled down to *extremely* low temperatures, don't resist electrical flow at all. They are essentially perfect conductors. The physics behind this phenomenon is fascinating and continues to dominate the attention of a good fraction of the community of condensed matter physicists. But for our purposes, what is important is that superconductors can carry *huge* currents. As we've seen, electric currents give rise to magnetic fields, and these can be enormous. For an accelerator, there is a balance. The bigger the magnetic field, the smaller one can make the accelerator's ring.

On the other hand, there is a big power bill involved in running superconducting magnets—they must all be cooled to extremely low temperatures, and the refrigeration required is expensive. So the SSC machine size represented an optimum balance of cost and performance.

From the late 1980s to the early nineties, progress on the SSC was rapid. Prototypes of the magnets were built and successfully tested, and plans for industrial-scale production developed. Digging of the huge tunnel to house the accelerator had begun. The big particle detectors required for the actual experiment went through extensive planning and were under development. Theorists and experimentalists worked on the problem of analyzing the vast amounts of data expected from the experiments. Theorists especially focused on making sure that a range of anticipated phenomena, from the Higgs to exotic possibilities like supersymmetry, would not be missed in the complicated experimental environment.

But there were other forces at work, which clouded the future of the SSC from the beginning. The overriding issue was one of cost. In round numbers, we are talking about an expenditure on the scale of $10 billion. This is a lot to ask the citizens of any country to pay for the pursuit of pure knowledge (even granting that there might be benefits from the work on the technologies required for the accelerator). Many states competed to host the machine, but once Texas was selected, there were forty-nine states with little financial interest in the accelerator (though efforts were made to distribute the contractors broadly). For a while, the project probably did benefit from the fact that President George H. W. Bush was from Texas. It suffered, however, from another problem, a sort of catch-22. Part of the original argument to Congress for the SSC was that this was an *American* project, which would enhance US scientific prestige. But as the budget issues grew more severe, the question became: Why aren't there more international collaborators? By the early 1990s, there was a new president (albeit of quite broad intellectual interests) and, more important, a political environment hostile to federal deficits. By 1993, support had weakened in Congress to the point that the project was canceled. Fans of *The West Wing* may remember (or want to watch) a 2002 episode inspired by the cancellation.

Through much of the SSC planning, fortunately, there was another proposal on the table: an accelerator to be built at CERN, the great European laboratory for particle physics, located in Geneva, Switzerland. CERN had been created in the years after the Second World War and played an important role in reinvigorating science and technology in Europe in those difficult days. A sequence of accelerators through the years made important discoveries at the laboratory. These included the W and Z bosons as well as precision tests of QCD and the theory of weak interactions. This latest proposed accelerator, the Large Hadron Collider, or LHC, would not have quite the capabilities of the SSC but, as originally conceived, was to be built quickly. While the SSC was an active project, many in the United States viewed the European proposal with a certain amount of disdain. Its projected energy and other capabilities were not the equal of the planned SSC. Perhaps, some American physicists thought, the Europeans were hoping to do a rush job, get in quickly to pick off low-hanging fruit, leaving the serious work for the SSC to do later. But with the cancellation of the SSC, all of that changed. My experimental colleagues at Santa Cruz had a major responsibility for one of the SSC detectors. Within hours after the SSC cancellation, they had been invited to join one of the LHC experiments, and they quickly retooled their efforts to focus on Geneva. I was rather awed by their spirit. I would have spent at least a few months mourning my wasted effort.

The LHC never faced quite the threat of cancellation that the SSC did, but it did encounter challenges, both technical and financial. What was imagined to take a few years took fifteen. Again, I was in awe of the patience and perseverance of my colleagues, and not just those at Santa Cruz. But gradually everything came together. The required magnets—also superconducting—were developed, tested, and manufactured (thousands of them in several countries). Twenty-seven kilometers (seventeen miles) of tunnels, with a depth at many points of over 500 feet, were dug. Two enormous detectors were built. In 2008, things were ready to go.

Not being an experimenter, I was not always on top of all the facts and figures about this remarkable device. I recall being part of a review of Fermilab in 2005. At some point, the director, in his remarks, spoke about the laboratory's extensive engagement with the LHC efforts, and mentioned that the energy stored in the machine would be equivalent to the kinetic energy of an aircraft carrier moving at forty knots (about forty-five miles per hour). I stopped paying attention—not a proper behavior in my position, but I couldn't help myself—to check this figure and quickly realized it was correct. I started to imagine a sci-fi movie, someone in a control tower as an aircraft carrier approached at almost highway speed, and the hero has to stop it. Pretty frightening.

After all the waiting, the LHC had a rocky start. Within just a few weeks, there was an accident, an electrical failure that caused some of the magnets to heat up, or "go normal" (cease to be superconducting). In the world of superconducting magnets, this is a catastrophic failure. The huge amount of energy stored in the currents in the magnets and in their magnetic fields is almost instantly turned into heat. This was exactly my sci-fi nightmare scenario, only the systems that control the machine had to deal with it in real time, figuring out how to dump the excess energy and save the machine. Fortunately, the LHC survived, but it had to be completely shut down. Hundreds of magnets had to be replaced and improvements made in the system to avoid a repeat of the disaster. This process took two years. Many fail-safe devices were put in place, and the machine was to be run for the time being at an energy barely half its design energy, since the original goal was no longer considered safe.

As the machine restarted, you can imagine how careful the accelerator operators were. But this time, things were different. The LHC performed beautifully. Over the next two years, hints appeared for a Higgs boson at an energy about 125 times the mc^2 of the proton. Finally, in 2012, on July 4, the evidence for the Higgs had reached the stage that it was a virtual statistical certainty, and the formal announcement of the discovery was made.

What did the experiments actually do? They studied literally trillions of collisions of protons. The energy of two protons when they meet in the LHC is about 8,000 times the mc^2 of a single proton—enough to produce 8,000 protons! In practice, several hundred particles of different types are produced in each collision. Two large detectors, ATLAS (for a Toroidal LHC Apparatus) and CMS (Compact Muon Solenoid), measured the properties of these particles as they emerged. These devices are huge (the ATLAS detector weights 7,000 metric tons, CMS 14,000 tons), on the one hand, and finely instrumented, on the other. They can determine properties of almost all of the particles that emerge—their energies, momenta, charges, and— critically—their identities. In fact, there were so many collisions every second (about a billion) and so many measurements made of each that it would be impossible to record all the information collected by the detectors. So computers made (and still make) decisions about which events are the most promising to reveal the Higgs or other new phenomena. Only about one in 10 million is recorded. That still left a data collection on the order of 15 petabytes (15 million gigabytes, about as much data as fit on half a million laptops, or about 1,000 times the amount of data in the Library of Congress). So how in this haystack do you find a Higgs? Only about one collision in 10 billion yields a Higgs particle. That number is so large that if you were to examine the products of one collision every second, you would see three events with a Higgs boson every thousand years. And when you have a Higgs particle, it doesn't come out to announce itself. The Higgs is very radioactive, meaning that when it's produced, it almost immediately undergoes radioactive decay, which is to say in about 10^{-25} seconds, or one millionth of a billionth of a billionth of a second. That's far too short even for the fastest electronics to see in any direct manner. Instead, it is the decay products that hold the clue. This is actually an application of the uncertainty principle. The limit on the ability to determine the energy of the Higgs particle—which shows up in the data—is related to the lifetime of the unstable particle.

Nature throws up further challenges to Higgs detection. Most of the Higgs decays are to a particular type of quark, the bottom quark (or b quark). But bottom quarks are produced in all sorts of other ways in accelerators, so many ways that, just accidentally, they will often resemble the decay products of the Higgs. Instead, the experiments first searched for a rare decay—the decay to a pair of very high-energy gamma rays. These, too, are produced in great numbers by many processes in accelerators, but these other processes can be accounted for accurately by theorists (itself an amazing tour de force). And it was known that for a Higgs of a 100 GeV or more, one could hope to see a small excess of these photon pairs. Indeed, this is where the two experiments found the first evidence of the Higgs particle.

Since the first announcement, the evidence has gotten better. Other decays of the Higgs have been observed, and all are consistent with the expectations of the simplest version of the Standard Model. We may know all the laws that govern nature down to scales 10,000 times smaller than an atomic nucleus. Further experimental studies, at the LHC and at machines under consideration in Japan, China, and CERN, will, among other things, look for small discrepancies in the properties of the Higgs from those predicted by the Standard Model, but the study of the Standard Model by way of particle physics may be nearly finished.

CHAPTER 7

STARDOM

Some physicists become famous, and some, despite having made spectacular contributions, do not. Hans Bethe, whom we met in the previous chapter, was famous, widely recognized as one of the greatest theoretical physicists of the twentieth century (and the early twenty-first). Born in Germany in 1906, his talents were recognized at a young age. With a Jewish mother, he was vulnerable when the Nazis came to power in 1933. After spending some time in England and in Italy, he came to the United States in 1935, spending most of his career at Cornell University. We've already encountered some of his work on quantum electrodynamics. He made numerous contributions to our understanding of the physics of atoms and atomic nuclei. During World War II, he led the group of theorists working on the atomic bomb at Los Alamos. After the war, he was a prominent advocate for arms control and sensible energy policies. He continued his research well into his nineties, and I had the good fortune to attend seminars he delivered with passion and eloquence toward the end of his life, on subjects including astrophysics and neutrinos. For our understanding of the universe, Bethe took a particularly important step: He explained how stars work. Prior to the discovery of the atomic nucleus and the enormous energy involved in nuclear reactions, the sun and stars were a mystery. If powered by ordinary chemical reactions, stars would burn up all their fuel in thousands of years, yet they power on for billions of years. Nuclear reactions involve millions of times more energy than chemical ones, and it was realized that this could account for the difference. It was Bethe who, in 1938, put together the modern picture of the nuclear reactions that power the sun and other stars. For his contribution to our understanding of stars, Bethe was awarded the Nobel Prize in 1967.

From this starting point, astronomers could develop a picture of how stars form, burn, and die. Stars are born when a large cloud of hydrogen atoms collapses under its own weight into a much smaller space. The tremendous squeezing of the material by the gravitational force heats it to enormous temperatures. The interior becomes so hot that some of the hydrogen nuclei, when they collide, undergo nuclear reactions, producing large amounts of energy and heavier elements. The burning goes on for billions of years. Eventually, the fuel is depleted and the star, no longer supported against gravity by its hot interior, collapses. The final phase is different for different stars, depending on the mass of the star. Our sun will go through a period as a large *red giant* before collapsing to an object known as a *white dwarf,* a dense, earth-size, non-burning star. More massive stars collapse but then explode, as supernovae, spewing much of their material into space and leaving behind either a neutron star or a black hole. The spewed-out stuff eventually collapses into other stars as well as planets, providing the raw materials—carbon, nitrogen, oxygen, iron—necessary for life as we know it. All of this makes a nice story, and it is consistent, at least, with much data in astronomy. But most of the action takes place deep inside the star, where we can't see what is going on. Or can we? Here the not-quite-so-famous scientist John Bahcall enters the story.

In 2016, I attended the American Physical Society meeting in Salt Lake City. I had come to watch Ed Witten, a close friend who will figure in chapters to follow, receive the first APS Medal for Exceptional Achievement in Research, for which I had nominated him. But one of the most memorable moments at that meeting was a talk by Neta Bahcall, a professor of astronomy at Princeton University. Her talk was about her late husband, the theoretical astrophysicist and astronomer John Bahcall. It was a loving story of a marriage and it was spiced with frustration that John did not win the Nobel Prize, as many physicists believe he should have.

Bahcall had many accomplishments to his credit, but his most consequential emerged from a focus on these questions: Precisely how does our sun work? And how can we *experimentally* test our understanding? For the first question, Bahcall took the ideas of Bethe and others and developed them into a very detailed model, including predicting exactly how hot the sun is in its interior (15 million degrees Kelvin, or 1.5×10^7 degrees).

As for the second question, what we can see easily is the

solar surface, and there is some information we can extract from these observations. The predictions of models can be tested against the measured temperature, and activities like sunspots and what we might call sunquakes (helioseismology—the analog of the seismology we do on the earth—was actually a subject of interest to Bahcall). But we can't see into the interior of the sun, much less any other, more faraway star, to verify our picture. The solar interior is extremely hot and dense, and the light itself—the energetic photons—can't stream out. Instead, photons produced deep inside the sun collide with electrons and nuclei with great frequency. The photons execute a sort of drunkard's walk, emerging from the sun only millions of years after they are produced in the core. By the time they begin their travels through space toward earth, their energies are only a tiny fraction of what they were when they were first produced.

But Bahcall recognized that there is another way to investigate the sun's core. Nuclear reactions in the sun and stars produce neutrinos, which are very different from photons. They interact extremely weakly with ordinary matter. So Bahcall realized that almost all the neutrinos produced in the sun would escape. Many would strike the earth a mere eight minutes after they were produced. The questions, then, were: Could you detect some of them, and how many did you expect to "see"? With his model, now called the *standard solar model*, Bahcall soon had an estimate of the expected number of neutrinos coming from the sun every second, and their energies. So how to detect them? Bahcall reasoned that if one built a detector with a lot of material, most of the neutrinos would pass through, but occasionally a neutrino would stop. Think of the material as a collection of atoms—more important, of atomic nuclei. A neutrino striking a chlorine nucleus can work a bit of alchemy. Chlorine, which became the principal target for the experiments, has 17 protons and 20 neutrons. The neutrino turns one of the neutrons into a proton and an electron; the resulting argon nucleus has 18 protons and 19 neutrons. The argon is radioactive; it decays with a half-life of thirty-five days.

Bahcall enlisted as a collaborator Ray Davis, a chemist at Brookhaven National Laboratory. Davis designed a detector, a large tank filled with cleaning fluid (carbon tetrachloride, a molecule with one carbon atom and four chlorine atoms). He developed a protocol to flush the tank every few weeks and watch for decaying argon nuclei.

Radioactivity is frightening, but compared to other poisons, it is easy to detect. Radioactivity typically involves the emission of fast charged particles. Such particles, as they pass through matter, knock electrons off atoms. The resulting trail of ions can be observed with a range of devices. For radioactivity, probably the most famous is the Geiger counter. The LHC and its two detectors are versions of this instrument writ large—*extremely* large. For Davis, the point is that the radioactivity of the argon can be reliably detected.

Bahcall and Davis viewed this as a new type of observatory, which, rather than looking for light from the sun or stars, would look for neutrinos. They realized that the experiment would have to be performed deep underground. Otherwise, many effects—most important, cosmic rays from space— would mimic the rare neutrino events the two physicists were seeking to count. Davis installed his tank in the Homestake Mine, a gold mine in South Dakota 4,850 feet underground. The Davis-Bahcall campaign went on for many years. Davis consistently found a rate for neutrino interactions about three times smaller than Bahcall predicted. Many—including me— were skeptical that this measurement was telling us anything interesting. Perhaps Bahcall's calculations were not as reliable as he claimed. It turns out that if he had the temperature of the solar interior wrong by a very small amount, the error could account for the discrepancy. The experiment itself might have issues.

But Bahcall kept checking and refining his computations and insisted that uncertainties in these could not account for the deficit of neutrinos. Davis performed many checks on the experiments, and others outside his group of collaborators examined their techniques and could not point to a systematic problem. Over time, interest in the question grew, and other experiments were proposed and carried out. These experiments were sensitive to neutrinos of different energy than the Davis experiment and, as a result, sensitive to different processes in the sun. This would help eliminate possible systematic concerns.

Meanwhile, a number of physicists began to consider another possibility. Perhaps the problem with the Davis experiments was not our understanding of the sun but our understanding of neutrinos. Neutrinos come in three varieties. There is a neutrino that is a partner of the electron, v_e; one that is a partner of the muon, v_μ; and one that is a partner of the tau. All of them are known to be quite light, far lighter than the electron.

At the time of the Davis experiments, many physicists assumed that, like the photon, neutrinos had no mass at all. But if they have a small mass, quantum mechanics allows an interesting possibility. Just as quantum mechanics doesn't permit one to say that a particle is simultaneously at a particular point and has a particular velocity, so then, if the neutrinos have a mass, one can't say with certainty which of the three types of neutrinos one has. Perhaps on their trip from the sun, the neutrinos, which start out as partners of the electron, might turn into neutrinos of the other types, which would not have been detected in Davis's experiment.

For skeptics like me, this was a lot to swallow. Not that neutrino mass itself (and "oscillations") was surprising. But the masses had to be in just the right range that oscillations happened as the neutrinos traveled to earth, not on much longer or shorter distances. It was as if some power was conspiring to spoil the Davis experiment. Yet the other experiments inspired by Davis's results also found discrepancies, which could perhaps be explained by neutrino mass as well.

But at this point, an anomaly emerged in another phenomenon involving neutrinos. In addition to neutrinos from the sun, neutrinos arrive on earth from cosmic rays. Cosmic rays consist of high-energy particles produced in deep space (many from other galaxies). The particles are mostly protons but also include heavier nuclei and photons. The nuclei, when they strike the upper atmosphere, produce violent nuclear reactions, and among the products are neutrinos. These neutrinos are mostly of the muon type, and they have much more energy than the neutrinos from the sun.

A different class of neutrino experiments studied the flux of these *atmospheric neutrinos*. Among them were the Super-Kamiokande experiment, located in a mine under Mount Ikeno in Japan. This detector found a deficit in the number of muon neutrinos! Again, for a skeptic, this seemed *really* too much to swallow. Now the length scale was much different, 100 miles or so, rather than 100 million; different neutrinos, different energies. So now one seemed to require two conspiracies. Much easier to imagine that we did not properly understand the sun and that physicists had not calculated correctly the number of neutrinos one expects to see from cosmic rays. But over time, with further experiments, the support for neutrino mass and neutrino oscillations became stronger. These included two other sets of experiments in Japan. One, called Kamland, studied neutrinos from the nuclear reactors around the country. The inner workings of reactors are well understood, and the number of neutrinos and the energies of these neutrinos emerging from the reactors can be reliably calculated. The results supported the oscillation hypothesis (the tragedy of Fukishima actually provided an important data point; the removal of that reactor from the data set modified—in an expected way—the overall Japanese neutrino flux). Another experiment aimed a beam of neutrinos produced in an accelerator near Tokyo, called KEK, at the Kamiokande mine. The experiment was called K2K. Again, the results supported the hypothesis that neutrinos have mass.

The clincher came from an experiment in Canada, known as SNO, for Sudbury Neutrino Observatory. This experiment was housed in a very deep nickel mine. It used a different strategy. All the other experiments looked for events where, in the neutrino interaction, a particle changed charge. For example, a proton turned into a neutron, while the neutrino turned into an electron. But there is another class of neutrino interactions, referred to as neutral current processes, where neither the neutrinos nor other particles change their identity. Detecting these reactions poses special challenges, but what is critical is that neutrinos of each type interact at the same rate. So even if neutrinos oscillate along the way, every neutrino emerging from the sun (or the upper atmosphere) should induce the same number of interactions. Therefore the total number of neutrinos detected should be as Bahcall originally predicted. Amazingly, the SNO experiment found precise agreement.

At this point, we have good measurements of many of the neutrino properties. Most recently, important results have been reported by an international collaboration (including China and the United States) working near the Daya Bay reactor in China (the bay faces the South China Sea). No doubt this line of research will continue and be a major part of the US Department of Energy's High Energy Physics program.

The story of the collaboration of Bahcall and Davis is a remarkable saga of persistent, careful work, of refusing to take the easy way out, and of insisting that one take surprising results seriously. Ray Davis won the Nobel Prize in 2002. But Bahcall was not included. Many have speculated as to the reasons, but the inner workings of the Nobel Committees are not public. On another personal note, shortly after the Davis prize,

I visited the Institute for Advanced Study. One evening, I was sitting in a public space with my daughter, Shifrah, then eight years old, and John came by.

We had a pleasant conversation about science, our families, and other matters. Afterward, I told Shifrah the story of Bahcall and Davis, and about the prize. She was outraged as only an eight-year-old can be. Later, as an undergraduate physics major, she wrote a paper on the discovery of neutrino mass. Traveling to work one morning with my car pool, I heard the news of the Nobel Prize announcement for the discovery of neutrino oscillations by the Super-Kamiokande and SNO collaborations in 2015 (specifically to the collaboration leaders, Takaaki Kajita in Japan and Arthur B. McDonald in Canada). I quickly pulled out my laptop and used Shifrah's paper, with its figures and graphs, to prepare a lecture for my class explaining the significance of the prize. Bahcall passed away in 2005. He was honored in various ways that included, I like to think, my daughter's ongoing interest in neutrinos. Still, the lack of Nobel Prize recognition clearly caused pain to Neta Bahcall and to John's many friends.

The question of how the neutrinos get their mass points to scales of length far smaller—probably 100 trillion times, or fourteen powers of 10—than an atomic nucleus. But like so many questions about the vanishingly small, they also promise to illuminate some cosmically large phenomena. We're headed now into the next part of the book. So far, we've focused on questions to which we know, or think we know, the answers. We're about to peer into less well charted territory—areas where we have burning questions, and speculations, some of which are plausible and subject to experimental test. As we'll see in the next chapter, neutrinos (and the processes that give them mass) could well be responsible for the existence of matter as we know it. If so, this happened extremely early, 10^{-37} seconds (about a trillionth of a trillionth of a trillionth of a second) after the big bang. For other questions, the science is profoundly unknown.

THE NEXT STEPS

CHAPTER 8

WHY IS THERE SOMETHING RATHER THAN NOTHING?

Dirac, we've seen, stumbled upon antimatter when he attempted to write down a quantum mechanical theory of the electron consistent with Einstein's principles of special relativity. The particle he predicted had a mass exactly equal to that of the electron, but with opposite electric charge. It soon became clear that this is a very general result. For every particle of a given charge, there is an antiparticle of the opposite charge, but with exactly the same mass. The antielectron—the positron—was discovered almost simultaneously with Dirac's theoretical musing. The antiproton was discovered with the development of large accelerators, in Berkeley in 1955. Many more such particles followed. Even various neutral particles have antiparticles—the antineutron and the antineutrinos for example. While the antineutron has the same electric charge as the neutron, it is distinguished from the neutron by its decays. Remember that the neutron decays to a proton, an electron, and an antineutrino; the antineutron decays to an antiproton, a positron, and a neutrino. An antineutron can collide with a neutron, annihilating to other forms of matter and energy.

By the end of the twentieth century, antimatter was a commonplace. Antiparticles of all sorts of particles were produced in accelerators. Beams of antielectrons colliding with beams of electrons were used in experiments at Stanford and CERN, where they were used to study particles like the Z^0. At CERN and Fermilab, beams of antiprotons colliding with beams of protons were critical to the discoveries of the Ws, Z^0, and the top quark.

The world of our experience, on the other hand, seems to be made up almost entirely of matter rather than antimatter. That's clearly good for us; colliding with antimatter, in the form of rocks, nearby planets or stars, or even not so distant galaxies would be truly cataclysmic. But maybe, in the universe at large, it's not this way. Of course, this is not a question many of us wake up each day worried about. But knowing a bit about antimatter, it is a logical possibility. Suppose there were, say, antimatter stars in our galaxy. We couldn't tell this easily by looking at them; the light they would emit would be quite similar to—in fact identical to—that from ordinary stars. But from time to time, a matter star would collide with (or even pass near) an antimatter star and the resulting explosion would be very impressive. Even passing through a cloud of matter dust (dust is what astronomers call hydrogen and similar atoms moving freely in space) would generate large numbers of high-energy gamma rays. These kinds of considerations allow astrophysicists to set limits on the amount of antimatter there might be in the observable universe, and the answer is that there can't be much.

We can, in fact, characterize the amount of matter in the universe today by comparing the number of protons plus neutrons with the number of photons in the cosmic microwave radiation. In a cubic meter of the universe, there are, on average, about half a billion microwave photons. In that same volume, you have less than a 10 percent chance of finding a proton or neutron. In other words, you have a chance of about 1 in 10 billion of finding a proton or neutron (a "baryon") for every photon you find. That's much worse than your chance of winning the jackpot lottery in your state. This number, 10^{-10}, is actually well measured and is known as the "baryon per photon ratio." It turns out that this number doesn't change with time, at least after some very early time. The shortage of matter relative to photons cries out for some explanation. If one were just making a wild guess, one might say that there should be comparable numbers of baryons and photons, or perhaps that there shouldn't be baryons at all.

Very shortly after the big bang, within our standard cosmology, the universe was far hotter than it is today. Remember that the higher the temperature, the more energetic the particles. When hot enough, a typical particle had so much energy that, by $E=mc^2$, it could turn some of that energy (E) into matter (with mass m), that is, a proton and an antiproton, or a neutron and an antineutron. This was the case when the universe was about one ten-thousandth of a second (10^{-4} seconds) old. So the number of particles and antiparticles would be about the same, and about the same as the number of photons. If we now run our clock forward, if there were exactly the same number of baryons as antibaryons, they would all annihilate, leaving behind just photons. There would be no protons and neutrons with which to build stars, galaxies, planets, people. Instead, to have a universe like the one we have today, there must have been a tiny excess of matter over antimatter, just right to leave over the number of protons, neutrons, and electrons we see now, about one for every 10 billion microwave photons.

Thought of in this way, the universe really contains a tiny amount of matter, and almost no antimatter. Cosmologists refer to this small excess of matter as the matter-antimatter asymmetry, and it is another very curious manifestation of the strange powers of 10 that appear in nature. Here we say that the ratio of the number of baryons—protons plus neutrons—to the number of photons is about 10^{-10}.

To appreciate how bizarre this is, we need to return to the symmetries of the laws of nature. There are two important ones in play here. First, there is a symmetry between particles and antiparticles—they have exactly the same mass and exactly the opposite charges. This follows, it turns out, from basic principles of quantum mechanics and Einstein's special relativity. If particles and antiparticles were truly the same in every way, it would be hard to offer any scientific explanation of why there are more particles than antiparticles in the universe. We would have to suppose that the universe was simply created that way, perhaps by some higher power. This power provided just the right amount of matter to explain why the universe is hospitable to people. Even scientists who subscribe to one or another of the world's religions tend to be uncomfortable with this as a mode of explanation.

Physicists give any would-be exact symmetry between particles and antiparticles a name—CP. By general principles of quantum mechanics and special relativity, this is equivalent to time reversal invariance, which we have discussed. But while particles and their antiparticles must have exactly the same masses, it turns out that the ways in which they react with other particles—the frequency with which they collide, for example—don't have to be exactly the same. The Standard Model *almost* has an exact symmetry between particles and antiparticles. We saw, when we learned about the weak interactions, that some slight differences have been observed experimentally, and that critical to explaining these is the third generation of quarks and leptons—the bottom and top quarks, the tau lepton and its associated neutrino.

When I took chemistry in high school, we were taught about other conservation laws. One was conservation of mass. This was useful for the experiments we did for class, but, thanks to Einstein, we know this is not exact: Mass can be converted into other forms of energy, and vice versa. This more sophisticated version my teacher called *conservation of matter.* In nuclear reactions, protons can turn into neutrons, and vice versa. But the total number of protons plus neutrons is always the same in any nuclear reaction. Once we allow for antiprotons and antineutrons, we have to generalize this. A pair of photons, for example, can collide and produce a proton and an antiproton. So we might insist that the total number of particles (protons plus neutrons) *minus* antiparticles (antiprotons plus antineutrons) must be the same before and after such collisions. Physicists speak of *conservation of baryon number.* This rule is almost exact in the Standard Model. If it's exact in nature, again, if we start with a certain baryon number in the early universe, it won't change. If it starts out zero, it will remain zero. We are in danger of having to invoke a deity again.

Andrei Sakharov was an important scientist and leading dissident voice in the former Soviet Union. Born in 1921, in the 1950s he played a critical role in the development of the Soviet hydrogen bomb. In the 1960s he emerged as an important critic of the communist regime. Of the Soviet Union, he wrote, "Our state is similar to a cancer cell, with its messianism and expansionism, its totalitarian suppression of dissent, the authoritarian structure of power, with a total absence of public control in the most important decisions in domestic and foreign policy, a closed society that does not inform its citizens of anything substantial, closed to the outside world, without freedom of travel or the exchange of information." For his work as an advocate for human rights and arms control, he won the Nobel Peace Prize in 1975. In 1980, he was sent into internal exile, passing away in 1988.

Back in 1965, shortly after the first discovery of the violation of time reversal (CP) symmetry, Sakharov realized that this might be an important clue as to how matter was created in the universe. He laid out three requirements on the underlying laws of nature in order that a universe initially with equal amounts of matter and antimatter would develop an asymmetry between them. Two of these requirements we've now encountered: violations of CP conservation and baryon number conservation. The third has to do with time itself: It must have an arrow. There must be a clear distinction between time moving forward and time moving backward. In our day-to-day life, the arrow of time is something we take for granted, if perhaps in a melancholy way. We grow older, and our bodies gradually wear out. The devices in our homes and workplaces age and cease to work as well. It's hard to imagine a time-reversed swimmer diving out of a pool and landing feetfirst on the pool's diving board, but such an event would obey Newton's laws. The distinction between the forward-going and backward-going events lies in the complexity of the starting and ending points. To set up the time reverse of the dive, we would have to push out the water droplets in the reverse direction, generating enough pressure to pop the diver out of the pool and back onto the board. This would be unimaginably difficult to achieve.

Entropy is a way of characterizing such complexity. In any "reasonable" situation, the entropy—complexity—grows. So time's arrow has something to do with the growth of complexity, or entropy. The state of the diver and the water is much more complex after the dive than before. Just how this analogy of popping out of the swimming pool might be realized in the very early universe is part of the question of how Sakharov's three conditions might have been realized just after the big bang.

What might violations of baryon number look like? A reaction that might violate baryon number is the decay of the proton to a positron and one of the mesons we encountered in the chapter on nuclear physics, the π^0. We can denote this decay symbolically as $p \rightarrow \pi^0 + e^+$. This obeys all the conservation laws we believe must be absolute—electric charge, energy, momentum, and angular momentum, but it violates baryon number. The initial proton carries one unit of baryon number, but the π^0 has zero baryon number (it consists of a quark and an antiquark, which are like 1/3 of a baryon and 1/3 of an antibaryon, respectively, so there is no net baryon number). The positron also has zero baryon number, so overall the decay violates baryon number. In fact, the π^0 is itself radioactive, turning into a pair of photons with a half-life of 10^{-16} seconds. So if this happens, the proton ultimately turns into a positron and two photons, leaving nothing to build the nuclei of atoms.

The attentive reader may realize that these reactions would violate another conservation law. In the decay process, not only does a baryon disappear but a lepton—an electron or positron—makes an appearance. Just as the Standard Model preserves the number of baryons minus antibaryons, it preserves the number of leptons minus antileptons—the lepton number. As we'll see, there is good reason to think that this conservation law must be violated in nature.

At the time Sakharov presented this idea, it was rather radical. Even today, despite much searching, there is no evidence for interactions that violate baryon number. If it happens, it's so rare that in a tank of water the size of a good-size house, it occurs less than once per year. So if such processes account for the observed matter-antimatter asymmetry, they must be more common in the early universe than at present.

The need for CP (time reversal) violation is also easy to understand. In the very hot early universe, the baryon number violating decay processes can also happen in reverse. A positron can collide with a π^0 particle, producing a proton, or an electron can collide with a π^0 and produce an antiproton. If these processes happen just as often (i.e., are equally probable), then if we start out with equal numbers of protons and antiprotons, we will always have equal numbers of protons and antiprotons. So processes involving particles and antiparticles *must not* occur at exactly the same rates. As we've said, time reversal, or CP, says that particles act exactly like antiparticles, and these processes *would* occur just as often unless the symmetry is violated. So CP violation is essential for there to be at least the possibility of creating an excess of protons over antiprotons.

Finally, Sakharov asserted that time must have a *preferred* direction. Of course, we humans view ourselves as slaves to the direction of time. We aren't able to "go back" and correct our mistakes and take advantage of missed opportunities (or dive backward out of swimming pools). For elementary particles, though, this is in some ways the most subtle of Sakharov's conditions. Again, we can consider the processes of proton and antiproton decays, and the reverse processes that create new protons and antiprotons. If there is no particular direction of time, these creation processes would happen as often as the decays. The baryon number would not change. Chemists and physicists describe a situation where a system is hot and reactions proceed backward and forward equally fast as being in thermal equilibrium. Feynman had a concise way of defining equilibrium; he said (roughly) that "a system is in equilibrium if all of the fast things have already happened, and all of the slow things have not happened yet." The sun provides an illustration of Feynman's definition. The sun currently is hot and is, at any given moment, to a good approximation in thermal equilibrium. Here, Feynman's fast things are the collisions of hydrogen in the center of the sun, or the scattering of light near the surface. The nuclear collisions in the center take place on time scales of a tiny fraction of a second. Photons produced in the center diffuse slowly to the surface through frequent, repeated scatterings of electrons and protons. It takes a few million years for radiation produced in the solar interior to reach the surface, and nothing much changes in the sun in that time. But the sun is slowly burning its nuclear fuel. In about 5 billion years, it will burn up so much of its fuel that it will drastically change, first collapsing and then growing to be a red giant. At this point, the sun will establish a different equilibrium for a while. This is an extreme example of Feynman's slow things.

In the story of the sun, we have an example of an arrow of time. There is a directionality in the flow of time in the process (with sad parallels in our own lives). As we've mentioned, this is closely connected with the notion of entropy, which can be described as the level of disorder in a system. A system in thermal equilibrium is very disordered. If one can't make an inventory of every atom and photon—and one can't—one can provide only a coarse description. One can say something about the sun's temperature, density, and size, but not too much else. The notorious second law of thermodynamics says that entropy never decreases with time; in practice, it almost always increases.

This condition, in the big bang cosmology, would seem easy

to satisfy. There clearly is an arrow of time; the universe is expanding as time goes by. We can even think of the size of the universe as a clock. Rather than state that the universe is 100 million years old, say, we can give its size at that time. But there is a problem. When the universe was 3 minutes old—very young compared to its current age—it was doubling in size roughly every 15 minutes. This may seem dramatically fast, but it is an extremely long time compared to the time between collisions of protons and neutrons. So, from the perspective of the elementary particles that make up the hot plasma of the early universe, the expansion is extremely slow. In Feynman's phrasing, the system was always (almost) in equilibrium. Something more drastic must happen earlier if the big bang is to give rise to the baryon asymmetry.

Sakharov put forward a model that would satisfy all these conditions. While providing a proof of principle that the matter-antimatter asymmetry might have resulted from microscopic processes in the early universe, it was not particularly plausible (and at that time, the major ingredients of the Standard Model were not even known). But we know much more about the laws of nature today, and we also have ideas about how the interactions of the Standard Model might be embedded in some larger theoretical structure.

Although Sakharov's model was not particularly convincing, it turned the question of the origin of matter into a scientific one, with striking implications. In particular, if the proton can decay, then *everything* is radioactive. If the half-life for the proton was like that of the neutron, we would all disappear in a few minutes. So that can't be. In fact, if the half-life was anything shorter than about 10^{16} years, there would be about 100,000 such decays in our bodies every second, and these would induce cancers that would soon kill us all. (This point was made by the physicist Maurice Goldhaber, who said that "we know in our bones" that the proton is very long lived.) So how long might the lifetime of the proton be?

Baryogenesis in the Standard Model

The first thing to resolve is: Does the Standard Model itself, along with the big bang, realize Sakharov's three conditions—violations of CP conservation, violations of baryon conservation, and time's arrow? For a long while, well into the 1980s, it was thought the answer was a firm no. But the issue turns out to be subtle. The Standard Model does violate CP invariance. It also turns out to have the possibility for a significant departure from equilibrium—for an arrow of time of the needed sort. As we have seen, the Higgs field is responsible now for the masses of most of the elementary particles. But the Higgs mechanism was not operative when the universe was very hot, hot enough that Higgs particles were copiously produced in collisions of photons, quarks, leptons, and other particles. At these high temperatures, the Ws, Z^0, and the quarks and leptons were without mass. The Higgs mechanism turned on when the temperature of the universe had cooled to about 10^{16} degrees Kelvin, or about 10^{-11} seconds after the big bang. When the Higgs mechanism turned on, so did the masses of the elementary particles. The transition between zero mass and nonzero mass can be sudden, and this turn-on establishes an arrow of time. For a Higgs as heavy as the one discovered at CERN, however, this turn-on is too slow to provide a meaningful direction of time. So the third of Sakharov's conditions is not satisfied.

At best, the Standard Model could be responsible for the matter-antimatter asymmetry in an alternative universe in which the Higgs particle is much lighter than it is in ours. But before we completely abandon it, let's consider the second of Sakharov's conditions: the violation of baryon number. One of the successes of the Standard Model is that it has a symmetry that preserves baryon number. This doesn't have to be built into the model; it's automatic, with no fudging. This isn't true of most proposals for physics beyond the Standard Model, as we'll see. Now, if the proton is radioactive, it has an enormous half-life. At a minimum, it must be long compared to the time that has elapsed compared to the big bang, or we wouldn't be here. So this is a triumph of the Standard Model.

That is not quite the whole story, though. Gerard 't Hooft, who played such an important role in our understanding of the weak interaction, discovered that there are subtle quantum mechanical effects through which the proton can decay within the Standard Model. He was able to calculate the proton lifetime due to this effect. It turns out to be an unimaginably long time. The lifetime is so long that, at least since the first second or so since the big bang, the chances that a single proton has decayed this way in the history of the universe is something like your chance of winning the big jackpot in a state lottery twenty times in a row. But others realized that these effects become larger as the temperature grows, and baryon number violation would be rapid in the very early universe. So the real obstacles in the Standard Model itself are twofold. First, the large mass of the Higgs, as we have already noted, and second, there is not enough CP violation to generate the observed asymmetry.

Grand Unification

The first compelling setting in which to explore the question of the stability of the proton and the production of the matter-antimatter asymmetry was provided by a proposal known as *grand unification*. This was an idea put forth originally by Sheldon Glashow, who we've also encountered in the story of the weak interactions, and Howard Georgi, also of Harvard University. While the Standard Model is very successful as a theoretical structure (already in 1974 one could place reasonable odds that the theory was on the right track), it seemed rather cumbersome. One feature that seemed unnecessarily complicated was the presence of three types of Yang-Mills interaction—one for the strong, one for the weak, and one for the electromagnetic interaction. Georgi and Glashow asked whether this structure might emerge from a single, larger gauge interaction. What was needed was a bigger symmetry. Combing math books, they found the simplest object that could incorporate all three. They built a model of particle physics based on this structure. In their picture, at extremely high energies, nature would exhibit a very high degree of symmetry. There would be not just the 12 gauge bosons of the Standard Model (the photon, the eight gluons, and the Ws and Z boson), but an additional 12 extremely heavy gauge bosons. The breaking of the symmetry and the large mass for these extra gauge bosons would result from a suitable implementation of the Higgs mechanism. These extra gauge bosons would carry electric charge (like the Ws) and also color. As a result, their interactions could change quarks into leptons—and violate baryon (and lepton) number. They would lead to decay of the proton in precisely the way we described above.

The next question is: What is the half-life of the proton predicted to be? This in turn depends on the mass of the heavy gauge bosons in a very sensitive way. If one increases the mass by a factor of 10, the lifetime becomes 10,000 times larger.

These extra gauge bosons would need to be extremely heavy. Knowing the strength of the strong, weak, and electromagnetic interactions, one can calculate the masses of these bosons, a calculation first done by Georgi, Helen Quinn, and Steven Weinberg. They found that the masses had to be about 10^{14} times the mass of the proton. Correspondingly, one could calculate the lifetime of the proton, about 10^{27} years. This is also an interesting number. In 100 kilograms of water, there are about 10^{29} protons. So if one could watch such a sample for a year, one would expect to see about 100 proton decays. If one had 50,000 kilograms (a modest swimming pool), one could hope to see several proton decays every hour!

Physicists quickly proposed experiments of just this type. To be successful, these experiments, like neutrino experiments, have to be deep underground so as to suppress backgrounds from cosmic rays. In addition to taking account of the small fraction of cosmic rays that penetrate deep into the earth, one has to worry also about fake events resulting from natural radioactivity. The first of these experiments was in the Soudan mine, an old iron mine in Minnesota. The experiment ran at a depth of about 2,300 feet. No proton decays were found.

Since that time, the prediction of the proton decay rate has become more refined, and current estimates are in the range of 10^{31} to 10^{33} years. More recent experiments, most notably the Super-Kamiokande experiment in Japan, 3,300 feet underground, have now achieved this level of sensitivity, and many theories have been ruled out.

We should give this whole process a name. Physicists refer to the creation of the baryon asymmetry of the universe as *baryogenesis*, for the creation of the baryon number. Grand unification provided the first well-motivated setting to consider this phenomenon. As for Sakharov's three conditions, we have seen that grand unified theories predict baryon number violation. They are built to incorporate CP violation, as they have to account for the CP violation observed in the Standard Model. So two conditions are pretty much automatic. What about the departure from equilibrium, the arrow of time?

The expansion of the universe provides an arrow of time; the universe is growing and cooling as time passes. But this needs to be tied in to what is going on more microscopically with the particles of the grand unified theory. For grand unified baryogenesis, the key players are the very heavy gauge bosons, often called X and Y. It is their interactions that violate baryon number. Because they are so heavy, their behavior is linked to time's arrow. The X and Y particles are *very* unstable, with half-lives of order 10^{-40} seconds. So they decay almost as soon as they form. When the universe is *extremely* hot, so hot that typical particle energies are larger than the mc² of the X and Y particles, reactions that produce these bosons occur frequently, and they are replaced as soon as they decay. But as the temperature gets lower, production stops, and the superheavy bosons decay and disappear. When one works out the details, one finds that one can readily obtain a baryon asymmetry comparable to what is observed.

Other Settings for Baryogenesis

Grand unified baryogenesis may well be the correct story. But there are some reasons to be skeptical. First, we have yet to see the smoking gun for grand unified theories—proton decay. Second, there are reasons to doubt that the universe was ever hot enough that there were lots of X and Y bosons. Inflation theory, which we'll encounter in chapter 12, is a successful model of the first fractions of a second after the big bang, in which the universe underwent a brief period of very rapid expansion. At the end of inflation, the universe's clock restarts, and it is unlikely that the universe was actually hot enough at this time to produce X and Y bosons.

There are other possibilities. One rather surprising one is connected with neutrino mass. As noted, the Standard Model, in addition to preserving baryon number, preserves lepton number. But neutrino masses almost certainly violate lepton number. In the very early universe, this violation of lepton number may be important. It can lead to the production of a net number of leptons. The effects that 't Hooft discovered violate baryon number also can turn leptons into baryons. So this is another way the baryon number can arise. This set of ideas is known as *leptogenesis*. It is particularly interesting because, over the next decade, we may acquire information of direct relevance from experimental studies of neutrinos. First, with a little luck, we may have some evidence as to how the violation of lepton number arises. Second, we might measure CP violation among the neutrinos. These would be only part of what we must know to be certain that baryons were produced this way, but they would provide some circumstantial evidence.

There is still another possibility. I have worked on both leptogenesis and grand unified baryogensis, but my personal favorite is another. There's a story I like to tell. When new graduate students come to my office, hoping to embark on a career in theoretical physics, I feel obligated to discourage them, at least a little bit. I explain that the field is very competitive and that their odds of finding positions in universities or national laboratories are not high. But I also explain that even if they do find a job, the chances of their making a major contribution to the field are also not high. I usually say that "you should be so fortunate as to be a footnote in the history of science." I then sometimes proudly point to a footnote in Andrei Sakharov's memoirs, where a mechanism I proposed along with Ian Affleck (who has subsequently become a noted condensed matter physicist at the University of British Columbia) is mentioned. The idea relies on a possible new symmetry of nature, called *supersymmetry,* which we'll encounter later. For now, the most striking feature of the mechanism is that it is extremely efficient, often producing *too many* baryons.

So physicists have several plausible ideas for how the asymmetry between matter and antimatter might arise. The question you should ask is: Will we ever know which, if any of these, is the correct explanation? At the moment, I can't say there is any way to test these ideas by looking in the sky. The evidence is more likely to be indirect. For grand unified baryogenesis, it could come from the discovery of proton decay. For leptogenesis, from further studies of neutrinos and their properties. For Affleck-Dine baryogenesis, from the discovery of supersymmetry. Time, hopefully, will tell precisely why there is something rather than nothing.

CHAPTER 9

"THE LARGE NUMBER PROBLEM"

In contemporary life, we routinely encounter huge numbers. In the political sphere, we speak of numbers like the budget of the US government (about \$5 trillion, or 5×10^{12} as I write). This is an extremely large number. If you sat down to count 5 trillion one-dollar bills, it would take you more than 10,000 years, assuming you counted one per second and never took a break to sleep or eat. Other huge numbers are the population of the earth (about 6 billion), and the number of internet searches per day (almost 6 billion, or one per person on earth).

In nature, there are also gigantic numbers at the high end of our powers of 10, and extremely tiny numbers, at the other end. The observable universe is about 13 billion light-years across. That's huge, corresponding to the large time that has elapsed since the big bang. The corresponding volume is also huge, particularly if viewed from the perspective of some of the smaller things we know in nature. In that volume, for example, one could fit about 10^{100} (a googol) neutrons.* A typical human is built of about 5×10^{25} atoms. A typical star has about 10^{55} atoms. We saw in the last chapter that the amount of matter in the universe is characterized by a very small number, about 10^{-10}, or one part in 10 billion. Our very existence is contingent on many of these numbers. The age of the universe, given that we are here to observe it, can't be much larger or smaller (in order of magnitude terms) than it is now. Galaxies and stars, much less people, would not have formed less than about a billion years after the big bang, and we've seen that the heavy elements needed to make us weren't present in the first generations of stars. The universe can't be too much older than it is now and support life.

Some of these numbers are related to some rather crazy numbers that appear in the laws of nature. The time it takes to form stars involves laws we don't know well but which determined the very early history of the universe. The size of stars, on the other hand, is something we can understand. Stars arise from a balance between gravity, pulling matter together, and the atomic forces between atoms (really ionized atoms—nuclei and electrons) pushing them apart. In the sun, the protons and electrons are separated from one another by about as much as they are in atoms. In an atom, the strength of the electric force between the proton and electron is about 10^{43} times the strength

---

*The name of the internet search giant is said to be a misspelling of this mathematical term, which was first introduced by the young nephew of the mathematician Edward Kasner, who did work early on on the possible cosmologies of Einstein's theory. The sorts of nerds who start such companies also speak of an even larger number, a *googolplex*.

of the gravitational force (more than a trillion trillion trillion times as strong). The volume of the sun is about 10^{56} times the volume of an atom. So we might suspect that, while they're not exactly the same, there is some relation between these huge numbers. And there is. Stars would be even larger if gravity was weaker than we observe it to be; they would be smaller if gravity was stronger.

Maybe the Constants of Nature Change with Time?

Paul Dirac, who played such an important role in establishing the quantum theory and predicted the existence of antimatter, was perhaps the first to note how puzzling some of these numbers are; he called this "the large number problem." While we are bothered by some numbers that are very small and some that are very large, we can think of very small numbers as very large, just by taking their reciprocals, so we can lump all such numbers together.

Dirac conjectured that some of these large numbers were related to the age of the universe. For example, the strength of the gravitational force might be getting smaller with time. The universe is now about 10^{38} times as old as it was when the temperature was just right for quarks to join into nuclei. So maybe at that time, the force of gravity between protons was about as strong as the electrical and nuclear forces, and it's gotten weaker in the intervening time by roughly the factor by which the universe has gotten older.

We have a lot of evidence against this explanation. First, we have good observational evidence that when the universe was much hotter—for temperatures at recombination when the universe was about 100,000 years old, or the temperatures of nucleosynthesis about 3 minutes after the big bang—the force of gravity had about the same strength it has now. Also, if Dirac's explanation was correct, the strength of other forces would be changing with time as well. But we have evidence that this is not the case. One dramatic and surprising bit of this evidence involves an event long ago in *earth's* history. In the mid-twentieth century, a uranium mine in Gabon, the Oklo Mine, was controlled by a French company. Because uranium can be used both for weapons and for power, the mining company was required to inventory just how much U^{235} it uncovered, and in 1972, it reported that there was slightly less than occurs normally. Concerned about possible diversion or theft, the French Atomic Energy Commission (Commissariat à l'Énergie Atomique, or CEA) launched an investigation and soon established that this depletion had occurred because, for a time, about 1.8 billion years ago, the uranium deposit in the mine had functioned as a naturally occurring nuclear reactor.

Freeman Dyson (mentioned in our discussion of quantum electrodynamics) realized that the Oklo phenomenon, while fascinating in itself, could be used to determine if certain numbers in the laws of physics were the same 1.7 billion years ago as they are today. Nuclear physicists are able to work out the details of these reactions so well that they can establish that there have been, at most, only extremely tiny changes to several constants of nature in the intervening time. So Dirac's idea of varying constants, while quite exciting, is largely ruled out.

Other Explanations for Large Numbers

There are several extremely large pure numbers among the constants that appear in the laws of nature. The electron mass is more than 100,000 (10^5) times smaller than the top quark mass. We don't have a good explanation for this. But much more extreme is the strength of gravity. Shortly after Max Planck put forward his quantum hypothesis, he realized that the strength of Newton's law of gravity could be translated into a mass or (through $E=mc^2$) into an energy. To be a bit more precise, we've spoken about the relative strength of the electrostatic attraction of an electron and proton in an atom for each other, and their gravitational attraction. If we replaced the electron and proton with much heavier particles, the gravitational force would be much stronger. Planck realized that a pair of particles would have a comparable gravitational as electric force between them, if the particles carried a certain standard charge and a mass 10^{19} times the mass of the proton. This enormous mass is called the *Planck mass*. It's a large mass compared to those of any of the elementary particles we know, but still not quite enough that you could feel the weight of one such particle if you held it in your hand. But you could feel the weight of a thousand of them.

This way of thinking about the gravitational force gives a way of thinking about Dirac's large number problem. The question is: Why is the Planck mass 10^{19} times larger than the proton mass? Or about 10^{17} times larger than the mass of the W, or Z, or the Higgs particle? These are crazy big numbers.

Actually, the first of these, from the perspective of our understanding of the strong force, turns out not to be weird at all. This traces to the fact that the size of the proton, while very small, emerges directly from the theory. By the uncertainty principle, this size is related to the mass of the proton.

But the Higgs mass, which in turn is related to the mass of the W and Z, is harder. The problem is that within the Standard Model, the Higgs particle is literally just a point, infinitely small. Correspondingly, the uncertainty principle would suggest that we don't know its speed—and thus its energy—at all. It sometimes will appear to have zero energy, sometimes enormous energy. What we observe should be an average of these two, something infinitely large.

One can, in a more precise way, try to calculate the effects of quantum mechanics on the mass of the Higgs particle, and, as our uncertainty principle argument suggests, one gets a nonsensical result. One can get something more reasonable if one assumes that the Higgs actually has a size, some sort of internal structure. Then the quantum corrections to the mass of the Higgs get larger as the size gets smaller. If the corrections to the mass are not to be larger than the mass itself, the size of the Higgs must be no smaller than about one-thousandth the size of the proton.

This particular example of Dirac's problem of large numbers—that the Higgs mass is so much smaller than the Planck mass—is called the hierarchy problem, or sometimes the naturalness problem. I first heard about this problem when I was a graduate student, in a talk by Leonard Susskind. In his talk, Leonard credited the physicist Ken Wilson, who we encountered in our discussion of the nuclear force, but Wilson actually never took this problem too seriously. Steven Weinberg did and considered similar possible solutions to those suggested by Susskind. Leonard formulated the problem in a particularly upsetting way. He noted that, while the mass of the Higgs gets larger as its size gets smaller, one can compensate by adjusting features of the theory carefully—more and more carefully as the size gets smaller. If the size was that suggested by Planck's formula—about 10^{-32} cm, then one has to cancel two numbers, not just in the first digit but in the second, the third, and so on—about 34 digits in all. While I've tried not to write too many formulas, allow me to write a hypothetical formula, just to indicate how absurd this is. One would need that the mass of the Higgs was the difference of two very similar numbers, something like

5,378,443,281,965,748,315,889,724,792,162,335,814
-5,378,443,281,965,748,315,889,724,792,162,335,262.

Susskind called this a problem of fine-tuning.

I remember being in total shock. I imagined some kind of omnipotent being trying to create a world like ours, turning dials ever so carefully. This seemed absurd. Leonard raised the consciousness of the community about this problem. He proposed a solution. Perhaps the Higgs particle is like the proton. It is a bound state of some particles held together by a force, which, by analogy to the "color" force that holds quarks together in nuclei, he called "technicolor" (if you're not of a certain age, you may not know that Technicolor was the name of an early process in which movies were produced in color rather than black and white. The journal to which he submitted his paper, the *Physical Review*, insisted he change the title to avoid trademark infringement). Instead of quarks, there were techniquarks, held together by analogs of gluons (which he called technigluons). Just as for the proton, the finite size of the bound object eliminates the problem of tuning or naturalness.

The idea is beautiful, and many people, myself included, devoted a great deal of effort to seeing whether it could work and what it would predict for experiments. Unfortunately, it rather quickly became clear that it was hard to construct models that agreed with known experimental facts about quarks and leptons. Still, the idea retains great appeal, and it has undergone revival in slightly different forms through the years, with names like "little Higgs" and "warped extra dimensions" (the subject of Lisa Randall's 2005 *Warped Passages*). At the LHC in CERN, much effort is going into the search for evidence of these phenomena, so far without positive results.

Returning to the problem of the Higgs mass, you might well be asking: What about other particles of the Standard Model? Aren't they points too? Don't they suffer from the same problem as the Higgs particle? The answers are yes and no. Susskind answered this question in his seminar. For the electron, for example, because of general principles, the terms that need to cancel to keep the mass small arise automatically. This can be traced to the fact that the theory has symmetries that fight against the mass.

In fact, in the early days of quantum mechanics, physicists asked exactly this question, and the answer wasn't so obvious. Before Feynman and others developed efficient techniques to consider the quantum theory of electromagnetism, Wolfgang Pauli assigned his student Victor Weisskopf the problem of calculating these contributions to the electron mass. When Weisskopf first did the calculation, he obtained a huge result. But another student pointed out to him that he had made a mistake. He had not taken proper account of a contribution coming from the fact that, due to the uncertainty principle, the electron can, for a brief interval, look like a pair of electrons and a positron. This extra contribution to the electron mass cancels others, leaving as a quantum correction only a small fraction of the original electron mass. Eventually, it was understood that this cancellation is the result of an underlying symmetry, a chiral symmetry similar to the one we have talked about for the strong interactions. Weisskopf went on to a distinguished career in theoretical physics, making important contributions to nuclear physics and eventually serving as director general of CERN. But his mistake—as important and instructive as it was—haunted him throughout his career and affected his choice of problems to tackle.

In the end, it is only the Higgs particle in the Standard Model that is vulnerable to this hierarchy problem. Surveying the situation, Gerard 't Hooft elevated this to a principle he called "naturalness." He argued that any theory of nature should be "natural" in the sense that large pure numbers should be accounted for by symmetries. Technicolor had this feature, but it failed to agree with experiment, at least without lots of ugly contortions.

Many theorists and experimentalists now moved in another direction.

Tempted by the Beautiful Mathematics of Supersymmetry

Susskind's technicolor idea, while clever, was basically a copy of physics we already knew. Perhaps something radically different was needed. Perhaps some new type of mathematics. For many of us, math can be frightening. It can also seem ugly in that it is hard and that it feels cold and removed from many of our day-to-day concerns. But some of us enjoy the challenges it provides and find it beautiful. Mathematicians and those who simply like the subject often see mathematics as a pursuit of absolute truths, not tied to human endeavors. Physics and mathematics have a complicated relationship. At least since Newton, one stimulates the other. Calculus proved a powerful tool to understand Newton's laws, and at the same time, the pursuit of the laws helped lead to the development of calculus as a crucial branch of mathematics. Among physicists, and especially among theoretical physicists, there are those who eschew fancy mathematics, feeling that it gets in the way of their struggles to understand the results of experiments, and there are those who love mathematics, and will take on problems more for their mathematical beauty than for their intrinsic scientific interest.

I have a foot in both camps. I want to understand phenomena—at the smallest and largest scales of distance. I view mathematics—especially contemporary mathematics—as quite difficult but sometimes a helpful tool along the way to questions that interest me. I have to confess, though, that I have on occasion been seduced by beautiful mathematics, sometimes precisely because it led to interesting physics.

But while I tend to straddle these two outlooks, the contemporary field of theoretical physics does sometimes have the appearance of two very different camps, which don't get along well together. History suggests, perhaps, that a balance between them is best. Einstein was hampered in his quest for his general theory of relativity by his ignorance, at the outset, of important developments in mathematics. David Hilbert, one of the great mathematicians of the period, had the mathematical tool but lacked the physical insight. Ultimately, under Einstein's influence, physicists learned a great deal of mathematics, and new ideas were injected into the world of mathematics by general relativity.

In the following century, theorists have drawn on many branches of mathematics, and sometimes developed new ones. One idea that has introduced beautiful mathematics and led to an extensive program of experiments is known as supersymmetry. Interest on the experimental side is driven by a possible role for supersymmetry in resolving the hierarchy problem. On the theoretical side, interest arose, at first, partly because this was a new mathematical structure, partly because of some hope that supersymmetry might have something to do with quantum theories of gravity. Over time, the mathematics has turned out to be exceedingly rich, yielding new insights in pure mathematics and in our understanding of physical theories. On the more experimental side, while the theory makes striking predictions, the story is, up to now, one of disappointment.

Supersymmetry and its relative, superstring theory, have

through the years been targets of attacks by physicists who worry that the direction of the field is too mathematical. I hope to make clear why the possibility that nature has this additional symmetry is compelling, and eventually what makes string theory so attractive. At the same time, in both cases, we'll see that current ideas are probably incomplete, and quite possibly wrong.

With 't Hooft's naturalness principle in mind, beginning in the early 1980s, some of us looked for a possible solution of the hierarchy problem in terms of a recently discovered, new type of symmetry, supersymmetry, recognizing that it might have the ability to solve the hierarchy problem in precisely the way 't Hooft prescribed. Supersymmetry is a curious type of symmetry, at that time rather unfamiliar. We have already spoken about isotopic symmetry and its generalization at the hands of Murray Gell-Mann. These symmetries are symmetries that relate different types of quarks. Supersymmetry is a hypothetical symmetry that, if it plays a role in nature, is in some ways similar but in some ways different. It would relate each of the particles we know—the quarks, leptons, photon, gluons, and so on—to new, yet undiscovered particles. What is striking is that the partners of each particle would carry a different spin. So, for example, the electron would be accompanied by a particle with no spin (like the Higgs) but with the same electric charge, called the selectron. The photon would be accompanied by a particle with spin like that of the electron but with no electric charge, called the photino (named in analogy to the neutrino). The quarks would be accompanied by spinless squarks, called squarks, and so on.

If supersymmetry were exact, each particle would have exactly the same mass as its superpartner. So, for example, the electron would have the same mass as the selectron. But this can't be the case. If it were, we'd find atoms with electrons replaced by selectrons, and protons and neutrons replaced by their superpartners. These would be pretty weird, since particles of spin zero don't obey the exclusion principle. As a result, we would have a very different sort of periodic table. So if supersymmetry is a symmetry of nature, it must be a broken symmetry. In fact, it must be pretty badly broken. And this might be just what we need to solve the hierarchy problem. Our main complaint about the Higgs boson is that its mass should be *very* large, since the Standard Model doesn't become more symmetric as we reduce the Higgs mass. As for the electron, with supersymmetry, there are different contributions to the mass of the Higgs, and with exact, unbroken supersymmetry, the contributions cancel out. The Higgs does not grow a large mass. If the symmetry is broken, the partners of the known particles have different masses, and the cancellation doesn't quite happen, but the corrections to the partner masses are smaller than the masses themselves. From the point of view of the fine-tuning we described, this argument makes a prediction: The mass of the superpartners should be something between the mass of the Z particle and about 10 times that amount, or about 1,000 times the mass of the proton. Our theories were not sufficiently precise or persuasive to predict just how much mass these particles should have, but the naturalness principle strongly suggested we would discover these particles at the LHC. Before describing what the experiments look for and the results, there are two other features of the supersymmetry hypothesis that are worthy of note.

Apart from predicting this raft of new particles and potentially solving the hierarchy problem, the supersymmetry hypothesis made two more dramatic predictions. Most of these new particles would be extremely radioactive. They would decay to more ordinary particles with a half-life ridiculously short, typically a trillionth of a trillionth of a second. That's so fast that, when produced in an accelerator, even if moving at nearly the speed of light, they would travel less than a trillionth of a centimeter—in other words, they would barely get past the point where they were created in the accelerator. But one of these particles must be different. It must be lighter than all the other new particles, and it must be stable. It must not carry electric charge, and in fact it should hardly interact with other particles, much like neutrinos. This extra-stable particle is known as the "lightest supersymmetric particle," or LSP.

There is a second prediction of this setup. The strength of the strong, weak, and electromagnetic forces are controlled by three numbers, called coupling constants, or just couplings. These numbers are independent; they are in the class of numbers in the back of a textbook, about which students—and most of the rest of us—don't ask too many questions. But if we assume that the forces, at some very short distance or high energy scale, are unified along the lines proposed by Georgi and Glashow, then, given the knowledge of the electromagnetic and weak couplings, the strength of the strong force is predicted. One gets a very different result with or without the supersymmetry hypothesis. In fact, when supersymmetry was first considered to resolve the large number problem, the couplings were not so well measured, and the supersymmetry version of the calculation did not agree with the experiments. That changed with more precise measurements at the Large Electron-Positron collider (the LEP), precursor of the LHC, where the agreement was impressive and so remains.

Two critical questions are: How much energy would an accelerator need to produce these particles? And how would these particles show themselves? The first question, how much energy is needed, is directly connected to the masses of these new particles (by $E=mc^2$). We have a rough idea what these masses should be. If we believe supersymmetry reduces or eliminates the hierarchy problem connected with the Higgs, these particles should have masses not too different from that of the Higgs itself. Perhaps we'll grant an order of magnitude or so one way or another. When the other particles decay, the debris—the decay products—always include one of the stable LSPs. Because the LSPs are so much like neutrinos, the experiments will almost never see them. So they just fly off, through the walls of the machine and beyond, carrying off energy with them. Thus, supersymmetric particles leave a distinct calling card: the production of some ordinary particles, and lots of missing energy. This is actually a very distinct experimental signature, and experiments have been searching for these particles at higher and higher energies from the time this hypothesis was put forward. We have also spoken of the dark matter. It turns out that this new, stable particle is a good—arguably ideal—candidate to fill the role.

If supersymmetry resolves the hierarchy problem, the masses of these particles are such that they should almost certainly be visible at the LHC. Because we don't know precisely the masses of any of these new particles, it is necessary to search, with as little prejudice as possible. Given the masses, though, we can calculate how many supersymmetric particles will be produced in proton collisions and how they decay. So far, the LHC experiments have seen no hint of supersymmetry, so they rule out certain masses for these particles. Physicists like to refer to these as "supersymmetry exclusions." The limits on the masses are becoming quite uncomfortable for supersymmetry proponents. Most (myself included) would have expected that if supersymmetry is there, we would have seen some sign by now.

None of the other ideas for understanding the hierarchy problem is faring better. It could be that some hint of new phenomena is just around the corner, but it could also be that our ideas about naturalness and hierarchies are just wrong. The LHC program still has a long way to go, gradually developing more intense beams and higher energies. We will see what the future holds.

For theorists, on the other hand, supersymmetry has turned out to be a bonanza of a different type. It has allowed them to understand questions about quantum field theories that are otherwise hard. Supersymmetric theories have special mathematical properties that make them particularly susceptible to analysis. This realization came about as some of us sought to understand the ways in which the symmetry, like the chiral symmetries of the Standard Model, can be broken. A path to address this question was opened by Ed Witten. Ian Affleck, Nathan Seiberg, and I were able to tackle the problem. In the course of this work, we realized that it was possible to address questions about these field theories with paper and pencil, which were, at best, accessible only in ordinary theories with supercomputers. Seiberg and Witten carried this much further, understanding, for example, mechanisms of quark confinement in some of these theories. This work has had enormous impact—the Seiberg-Witten papers have been cited thousands of times. Supersymmetry remains a particularly powerful tool to answer fundamental questions about what form new laws—I am tempted to say the fundamental laws—of nature may take.

While the results from the LHC suggest that physics that might resolve the hierarchy problem is not within reach, one experimental puzzle may indicate that there *are* some new phenomena just around the corner. In our discussion of QED, we spoke about the incredible agreement between the measured value of the electron magnetic moment and the theoretical prediction. One can ask—both experimentally and theoretically—the same question for the muon. Here, there is a slight discrepancy. The measured and theoretical results both are known at a similar level. The numbers agree to about a part in 10^{11}, i.e., amazingly well. But there is a persistent disagreement in the last decimal place, established in 2000 by an incredibly sensitive experiment at the Brookhaven National Laboratory. This number has attracted much attention. This discrepancy *might* be explained by the sorts of new particles expected for supersymmetry and other attempts to solve the hierarchy problem. The result was so dramatic that physicists at Fermilab, near Chicago, proposed a still more sensitive version, taking advantage of facilities at the lab to produce muons. The entire

Brookhaven apparatus was moved to Illinois, by road and river, and upgraded, and the experiment was run. Just as I was completing this book, the results were announced. Indeed, given the potential for human bias to impact the analysis yielding this single number, the data was "blinded." In other words, those studying the data could not tell exactly what number they had extracted. When the final number was revealed, it agreed well with the Brookhaven measurement, and the disagreement with the theoretical prediction became harder to shrug off. I personally am hopeful that this might represent new and exciting physics but am sufficiently conservative that I am devoting some effort to assess possible limitations of the Standard Model calculations. I don't expect the experimental result to move. Indeed, the experimental and theoretical efforts on this problem have been heroic. I am in awe of the level of scientific integrity displayed by the refusal of each side to fudge this tiny disagreement.

CHAPTER 10

WHAT IS THE UNIVERSE MADE OF?

I am a great admirer of astronomers, and one of the pleasures of working at UC Santa Cruz has been the opportunities it offers to become friends with an outstanding group of them. One of the best-known is my colleague Sandra Faber, who, in addition to many discoveries, was the prime mover of the rescue of the Hubble Space Telescope in 1993 and was awarded the National Medal of Science by President Barak Obama in 2013. Shortly after I first moved to Santa Cruz, Sandy and I attended a fancy dinner event in San Jose as appendages of our spouses (Sandy's husband is an attorney). Clearly at a bit of a loss as to what to do with us, the managers of the event hit on the idea of seating us together, and I have to say it was a great evening. We talked a lot about science. We also compared physics and astronomy as fields. I remember well Sandy saying that the reward of a career in astronomy is that one gets to ask the same questions one asks as a child through one's life, just with more and more sophistication. We live not far apart, and through the years, Sandy and I would sometimes carpool together. Sandy is extremely sharp and quick to see the central question in any endeavor, and I often found myself on the defensive, responding to questions like—Why is that interesting? Why are you working on that? Why are you interested in hiring so-and-so? But she was also an important source of support, and one summer was the research adviser to my younger daughter in a program for high school students.

One of those questions we might have asked in our childhood is: What is the universe made of? By the 1930s, scientists understood that the basic building blocks of the stuff around us are atoms, which are made of protons, neutrons, and electrons. From these, we developed an understanding of such relatively massive objects as stars. It was natural to assume that these particles accounted for pretty much everything in existence. This is not the case—most of the matter of the universe exists in some other form. This is the notorious *dark matter.*

Astronomers can make an inventory of the things they can see with telescopes of various types: mostly stars and what they call dust (primarily hydrogen gas), out to distances billions of light-years away from the Milky Way. Adding all of this up, one can have some idea of what the universe *weighs* (the answer, roughly, is 10^{52} kilograms—or pounds, it doesn't much matter). That's something like the mass of 10^{78} atoms.

But there are also objects in space that can't be seen directly with telescopes. In recent years, one of the most dramatic discoveries in science has been planets outside our solar system. For many of us, it is perhaps not such a big surprise that there are such planets. Why should our solar system be unique in this way? Other planets, perhaps some with intelligent life, have been the subject of science fiction for as long as the genre has existed. But how can astronomers possibly find such objects? They are not like stars, giving off their own light. At best, they reflect a little bit of light from their nearby suns. The nearest of these planets found so far, Proxima Centauri b, is 4 light-years from earth. That little bit of light has spread out over a huge area by the time it travels so far; what reaches earth is less than a billionth of the light from a star a few light-years away. There's no hope to detect it directly.

Instead, the strategy that has led to these discoveries—almost 4,000 planets to date—is to look for the effect of the planets on the stars they orbit. With sensitive instruments, astronomers search for small wobbles in the motion of these stars, and using Newton's laws, they can determine how heavy these planets are. Not surprisingly, the first discoveries involved large, Jupiter-size planets, since these have the biggest pull on their stellar companions, but recent discoveries have included earth-size planets. We've learned, if there was any doubt, that planets are common, and we will eventually have some idea how many planets there are which might support life.

I'm as interested as anyone in the possibility that there might be intelligent life elsewhere in the universe, and how close our nearest companions might be. But for our purposes here, the discovery of extrasolar planets illustrates that one can hunt for mass indirectly, by looking at the motion of stars and galaxies and inferring, using Newton's laws, how much stuff is pulling on them. For large objects like galaxies, these sorts of studies came early. An astronomer who pioneered such observations was Fritz Zwicky. Born in Bulgaria in 1898 and educated in Switzerland, Zwicky came to the United States in 1925, where he spent most of his career as an astronomer at Caltech. He played an important role in the discovery of neutron stars and supernovae, and he studied and cataloged galaxies. He had a reputation as a curmudgeon, but he was also a humanitarian.

In the 1930s, Zwicky studied the motion of stars in the Coma Cluster, a collection of more than 1,000 galaxies about 320 million light-years from earth. The cluster itself contains over 100 trillion stars. He found that the visible stars could not account for their rapid motion. Their gravity, by itself, would not be enough to keep them from flying apart. He hypothesized that there must be more matter in the cluster than can be seen with telescopes, and he gave it the name dark matter. While interesting, there was skepticism for a long time about the result, and it was not clear that this feature of the Coma Cluster was typical. In fact, Zwicky significantly underestimated the amount of normal matter in the cluster, and so overestimated the amount of dark matter.

Persuasive evidence that dark matter is typical in galaxies came only in the late 1970s, through the work of Vera Rubin and her collaborator, Kent Ford. Born in 1928, Rubin faced many challenges as a woman in science in a very sexist era. Neta Bahcall, who we encountered in the last chapter, was Rubin's colleague in Princeton. She tells several stories. One is of a department chair who suggested that he should present Rubin's research for her at an upcoming meeting even though she would, in fact, be present. She replied. "That's okay, I'll do it." For a long time, she was unable to use the observatory on Mount Palomar (for some time the largest existing reflecting telescope) as a woman. When she was finally able to observe in the 1960s, she took in stride the fact that there were restrooms only for men. She received many honors, including the naming of a major telescope project in her memory, and passed away in 2016. Among Rubin's legacies, we can count many exceptional students, including my colleague, Sandy Faber.

By now, the evidence for dark matter comes not only from motion of stars in galaxies but from motion of galaxies within clusters of galaxies and, more indirectly, from "lensing," the bending of light from stars and galaxies by mass on its way to the earth. From Einstein's theory we know that gravity alters the paths of light rays (photons) just as it does for more familiar objects with mass. Studying how images of stars are distorted on their way to earth, astronomers have found evidence of large amounts of invisible mass between the stars and earth. Further evidence for the dark matter comes from studies of the abundances of the elements in the universe, and the cosmic microwave background radiation.

We know, with some certainty, that there is about five times more dark matter than ordinary matter. Dark means just that: Whatever it is, it doesn't emit light. But we know more than that—or more precisely, we know what we don't know. We can say with some confidence that this matter is not in the form of, say, dark stars or uncounted planets; dedicated experiments have ruled out these possibilities. This conclusion is also supported by the indirect observations we have mentioned: the abundance of (light) elements and the features of the cosmic microwave background radiation each. So what is it? Almost certainly, it is some new kind of elementary particle. This particle must have mass—to be the dark matter, the typical particle must be moving slowly today. But it must not carry electric charge, otherwise we would—quite literally—see it. It would reflect and emit light. In fact, it must interact hardly at all with ordinary matter, except for its gravitational pull.

Maybe Supersymmetry Explains the Dark Matter?

One of the reasons supersymmetry has attracted so much interest among particle physicists is that it predicts—(almost) automatically—a candidate for the dark matter. Supersymmetric models require a host of new particles, at least one new particle for each known particle. In addition to the electron, we have the *selectron*, as mentioned above. For questions of dark matter, we are interested in the particles that are electrically neutral. These might be the partners of the photon (the *photino*), the Z boson (for rather technical but still whimsical reasons, the *bino* or *wino*), or the neutral Higgs particles (*higgsino*) that would carry spin, like the electron. Alternatively, they might be the partners of the neutrinos, the *sneutrinos*. These spinless particles would be electrically neutral and interact quite weakly with ordinary matter. But really important is that, in most hypotheses for how supersymmetry might be realized in nature, the lightest of these particles, the *lightest supersymmetric particle*, is absolutely stable; it is not at all radioactive. So if somehow these were produced in the early universe, and in precisely the right quantity, the lightest of the particles would be just what is needed to play the role of the dark matter.

More generally, heavy dark matter candidates of this type have been dubbed WIMPs, for Weakly Interacting Massive Particles. The term captures the notion that the particles are typically much heavier than the proton but interact with other types of elementary particles and with each other much more weakly; the interactions are typically as weak or even weaker than those of neutrinos. But now comes the truly remarkable point. In the early universe, without fiddling (much) with the basic model, these particles are produced in just the right quantity to account for the observed amount of dark matter.

As we have seen, at very early times, the universe was extremely hot. Even 100,000 years after the big bang, the temperature was about 10,000 degrees Kelvin. At earlier times, the temperature was higher still. Once again, it is convenient to think in powers of 10. When the universe was a factor of 10,000 times younger, i.e., ten years after the big bang, the temperature was about 100 times larger, or about 10 million degrees Kelvin. At this time, atoms collide with each other so violently that all their electrons are stripped off. The universe consists of a plasma of ions and electrons. At much earlier times—when the universe was younger by a factor of 10^{20} or so (100 billion billion!)—the typical energies of particles were enormous. The energies were about 1,000 times the mc^2 of the proton—typical particles had enough energy to produce 1,000 protons. That's also larger than the rest energy, mc^2, of the WIMPs. As a result, collisions of particles frequently produce WIMPs. These WIMP particles, in turn, can decay, or annihilate with others, producing high-energy gamma rays or other particles, about as many as photons, electrons, or other particles. So in this extremely hot plasma, there are lots of WIMPs.

The next question is: What happens as time goes by and the universe cools? Eventually, collisions of quarks and leptons don't have enough energy to produce WIMPs. But while the WIMPs may be stable, they can collide with other WIMPs, annihilating and producing other forms of energy. One can calculate how often these annihilations occur. It turns out that most of the WIMPs do annihilate, but the small fraction left over can readily be just the right number to account for the observed dark matter. In typical models, this means there are about 100 billion ordinary particles—quarks, gluons, electrons, photons—for every WIMP. (A hundred billion is roughly the number of people who have been born on the earth in its whole history.) How could we tell?

Detecting WIMPs

Edward Witten is one of the leading theoretical physicists of the last few decades. He is an extraordinary intellect. He is exceptionally smart, but also exceptionally hardworking and very disciplined. While mathematically more adept than most, he is also remarkable in focusing on simple, conceptual questions in physics.

I first encountered Edward when I was still an undergraduate student. In those days, I was studying physics and thinking about graduate school, with an emphasis on theoretical physics. Some of my teachers were quite discouraging. I remember one of them saying to me, "Only geniuses succeed in that field. You should look to other subfields of physics for a career." On a visit with my parents in Cincinnati, they mentioned that an old friend of theirs, himself a physicist at the University of Cincinnati, had a son who was interested in theoretical physics, who had just started graduate school in Princeton. Perhaps I wanted to meet him? So we had him for dinner. And by the time dinner was over, I was thoroughly depressed. While a very pleasant person, this physics student was so much smarter than me, and knew so much more. Clearly, my teachers were right. After all, if some random friend of my parents in this field was this smart, they must all be that way. What I would eventually come to appreciate was that this person—Ed Witten—was quite possibly the smartest person in the field. Foolishly, perhaps, I persisted in my determination to do theoretical physics. Edward, his spouse, the physicist Chiara Nappi, and I became good friends. From time to time in my career, I have managed to tell him something he found exciting, and we have had several delightful collaborations through the years. These often started with Ed criticizing some proposal or suggestion of mine. More than once, he took some half-baked suggestion of mine and turned it into a gem. At various times I collaborated on projects with Chiara as well.

Witten is sometimes criticized as being overly focused on mathematics. First, it should be said that to the extent that's true, his contributions to pure mathematics are remarkable. But Edward also has great physical insight. In the early 1980s, he was regularly bugging me and others that we should be thinking about the dark matter, what it might be and how one might detect it. I, for one, was distracted with other questions and, frankly, probably just a bit too lazy to pursue the suggestion. In 1984, with his graduate student, Mark W. Goodman, Witten wrote the seminal paper on the problem of detecting WIMPs in experiments. They started by noting that if such particles constitute the dark matter, they are all around us, passing through us, the earth, and our laboratories. On their way through a chunk of material, they can collide with electrons and atomic nuclei. These collisions are something like bowling balls hitting Ping-Pong balls, since the WIMPs are so heavy. The WIMPs are barely deflected, but they leave a little energy behind in the form of the recoiling nucleus. Goodman and Witten realized that while small, this little bit of energy could possibly be detected—the squashed Ping-Pong ball.

The paper of Goodman and Witten, as well as papers proposing how operationally such experiments would work, have led in the subsequent decades to experiments around the world designed for such direct searches for dark matter. The ideas for detecting these small deposits of energy are quite ingenious; the detectors have evolved from small prototypes to large and expensive devices, funded by various government agencies around the world.

Direct detection experiments confront a number of challenges. They are searching for tiny, infrequent depositions of energy. So, to start, it is necessary to shield their equipment from cosmic rays and background radiation. All these experiments, like the neutrino experiments we talked about earlier, are located deep underground. Second, the detectors need to be optimized for sensitivity to tiny amounts of energy. Different experiments have used different detection materials—silicon, germanium, and xenon, for example. Silicon is familiar—it is found in sand, and it is the basic material of our electronics technology. Germanium is a more expensive, close relative. Xenon, for those who remember their chemistry, is a noble gas. It is rare and quite expensive, but a powerful tool for these studies.

So far, with one possible exception, none of these well-made experiments have reported a discovery of dark matter. We do know that if the dark matter does consist of WIMPs, these particles are surprisingly heavy and interact much more weakly than expected with ordinary particles. Some of the most stringent limits to date on dark matter have been set by a detector called CDMS, for Cryogenic Dark Matter Search. A prototype ran under the Stanford University campus, but the experiment has now run for many years in the Soudan mine, a deep iron mine in northern Minnesota. CDMS, and its successor experiment, SuperCDMS, use detectors made from silicon and germanium, cooled to extremely low temperatures, a few thousandths of a degree Kelvin (this is the "cryogenic" part of the name). For proponents of supersymmetric dark matter, the failure to make a discovery has been a disappointment. Other experiments have produced similar exclusions.

One experiment, however, has for some time reported a possible signal. This experiment is a detector, situated in a highway tunnel under the Gran Sasso, the highest peak in the Apennine mountain range in Italy. This is a remarkable laboratory, created by the Italian government, recognizing that, with

1,400 meters of rock (about a mile) above it, experiments in the halls hollowed out of the mountain are well shielded from cosmic rays. The DAMA experiment uses a material called sodium iodide as its detector, and it looks not so much for individual dark matter particles as for different rates of production at different times of year. The point is that the rate at which dark matter particles interact is proportional to their velocity. At different times of year, the earth is sailing into, or away from, the cloud of dark matter. More precisely, on cosmic scales, the scales on which the universe looks boring—homogeneous and isotropic—the dark matter particles are, on average, at rest. But the earth in its orbit is moving relative to this cloud. So at different times of year, the detector should see more, or fewer, particles.

The experiment has been running since the 1990s and has seen a seasonal variation in its particle counts. But the results have been surrounded by much controversy, and another experiment, in the Gran Sasso tunnel as well, using xenon (and called Xenon) has contradictory results. Many theorists and experimentalists have put forward ideas to reconcile these experiments, but the situation remains unsettled. As time passes, it is becoming harder and harder to reconcile the DAMA result with the negative results of other, more sensitive, experiments.

There is another, quite different strategy for looking for dark matter. This is known as *indirect detection*. Dark matter is everywhere but is more concentrated around galaxies than in the spaces between. If the dark matter consists of WIMPs, from time to time one will collide with another and annihilate, just as matter and antimatter annihilate. These collisions will produce other forms of energy. Much of this would emerge as gamma ray photons, some of which would reach the earth. With suitable detectors on the ground or in space, we might hope to detect this radiation.

For a typical WIMP mass, the energy of these photons would be hundreds of millions of times larger than that of your dental X-rays. The most likely place for such events to originate is in regions of our own galaxy where the dark matter concentration is highest. This place is expected to be the center of the galaxy. Astronomers and astrophysicists have some idea of the concentration of dark matter near the galactic center, but there are large uncertainties, so the exact number of photons one expects to see for a given type of dark matter can be only roughly estimated. For many types of dark matter, however, there is the hope of seeing large numbers of such gamma rays.

Scientists have built a number of instruments that have the capability to detect high-energy gamma radiation from space. One of these is the FERMI satellite, launched from Cape Canaveral in 2008. The FERMI satellite can measure the energies and directions of photons from space with great precision. An earlier satellite, EGRET, had studied high-energy gamma rays from space, but FERMI is sensitive to a much broader range of energies and has greater capability to determine the energy and the direction of origin. One important part of the EGRET program was the study of gamma-ray bursts, the brightest electromagnetic events in the universe. FERMI has continued these studies, but its research program is much broader.

The story of the gamma-ray bursts itself is quite remarkable. These events were discovered not by astrophysicists or astronomers but by the US Department of Defense, at the height of the Cold War. In 1967, the United States launched the Vela satellites, designed to look for possible Soviet violations of the Nuclear Test Ban Treaty. Detectors on one of the satellites observed bursts of high-energy gamma rays. Initially, it was feared that these were signs of an impending nuclear attack, but that worry was quickly ruled out. From repeated observations, it was soon realized that signals seen by the satellites were due to previously unobserved phenomena in deep space. With the EGRET satellite, it was understood that these involved enormous bursts of energy. Truly sorting out what was going on, however, has required the FERMI satellite.

The detector on FERMI is very sensitive and capable, so it has been possible to study not only gamma-ray bursts but a variety of other astrophysical phenomena. It is also possible to do searches for dark matter. I can't help but boast that several of my Santa Cruz colleagues have played critical roles in the project from its inception. Bill Atwood and Robert Johnson developed the main instrument (along with Peter Michelson of Stanford). Steve Ritz was the scientific director of the project at NASA, before becoming director of the Santa Cruz Institute for Particle Physics (SCIPP). Hartmut Sadrozinski and Terry Schalk, two of my car pool buddies, played central roles from the beginning.

FERMI has made many discoveries. As far as dark matter is concerned, though from time to time there have been some tantalizing anomalies in the data, the observed gamma-ray signals on the satellite are consistent with those expected from known astrophysical processes. Other satellites also have been deployed around the world with missions including the hunt for dark matter. One, known as AMS (for Alpha Magnetic Spectrometer), looks for positrons (antielectrons) in space. The principal investigator of the experiment is Sam Ting, an elementary particle experimentalist who led one of the experiments that discovered the charmed quark. AMS has a number of goals. One is to establish whether or not faraway parts of the universe might consist of antimatter rather than matter; most theorists don't think this is a likely outcome. Another focus of the AMS research program is on the search for dark matter. Just as dark matter annihilation can produce pairs of photons, it can also produce pairs of electrons and positrons. From this and other experiments, there have been small anomalies in the data that *might* have something to do with dark matter. The question one always asks is, Can these discrepancies—for example, an excess of positrons in the AMS data—be due not to dark matter annihilation but to violent astrophysical phenomena? Resolving this is an area of intense activity.

Axion Dark Matter

The WIMP paradigm for dark matter emerged from thinking about the hierarchy problem and the resulting hypothesis that the laws of nature might exhibit a new symmetry: supersymmetry. Assuming that the resulting new particles accounted for the hierarchy problem, they had the right masses and interaction strength to account for the observed density of dark matter. All this was automatic. No fudging. A beautiful picture, if true, and this was one of the reasons the supersymmetry hypothesis had such traction for so long. But by now, supersymmetric particles of the expected masses are largely, if not completely, ruled out. Similarly, WIMPs of the expected masses have failed to show up in direct-detection searches. Before throwing up our hands in despair, there is another dark matter candidate—also motivated by one of the great questions about the laws of nature—with the right properties to be the dark matter. This particle, known as the *axion*, has been the subject of much attention for about as long as WIMPs.

In our discussion of the strong interactions, we have mentioned a problem known as the strong CP problem. As we saw, Newton's laws obey a symmetry called time reversal. Breaking of this symmetry, by a small amount, is crucial to the story of baryogenesis. But one of the well-studied features of the strong interactions is that they respect this symmetry, which, as we discussed, is connected to another symmetry, called CP.

The Harvard experimentalist Norman Ramsey realized in 1951 that, for the strong interactions, there is a very sensitive test of this property. The neutron, we've said, is very much like the proton. It has almost the same mass, but unlike the proton, it has no electric charge. As a result, one might think that nothing would happen to it in the electric field of Faraday and Maxwell. But even though it has no charge, it might have electric properties such that it could be pushed or pulled along in an electric field. After all, the neutron is built of quarks, and the quarks themselves have charges. So even though it's electrically neutral, it would be affected by electric fields, at least to the extent that the quarks are separated in the neutron by a little bit. Physicists characterize this in terms of an "electric dipole moment." Ramsey reasoned that the neutron might have a dipole moment of roughly its size, 10^{-13} cm. But he also realized that if time reversal is an exact symmetry, a dipole moment is forbidden. He found, in his first experiments, that if there is a moment, it is at least 10 million times smaller than the simple guess. Current experiments show that this moment, if it exists, is more than a trillion times smaller.

When the theory of strong interactions was first written down, one of its appealing features was that it seemed to explain this fact automatically. But Gerard 't Hooft, whom we remember for his role in the development of the weak interaction theory, pointed out that this is not the case. The reason is subtle; it involves a fancy bit of modern mathematics that, at the time, was not familiar to most physicists. 't Hooft explained that when we write the equations of the theory, there is one term we can add that does violate time reversal symmetry. This term is proportional to a number, usually denoted by the Greek letter θ (theta). If you were creating the universe and choosing this number at random, you would likely come up with something like 2.

Two of the great mathematicians of the twentieth century, Michael Atiyah, originally from Lebanon, who spent most of his career at Oxford University, and Isadore Manuel Singer of MIT contributed to elucidating this mathematics. In the 1970s, Singer had two PhD students, Daniel Friedan and Roger Schlafly. One was the son of noted feminist Betty Friedan (author of *The Feminine Mystique*), the other the son of the noted conservative activist Phyllis Schlafly (subject of the recent movie *Mrs. America*). Dan became a professor at Rutgers University and has been a leading figure in string theory. Dan and Roger received their PhDs the same year, and it seems that both mothers came to graduation. I have not been able to ascertain if there were fireworks.

In any case, given the unfamiliar mathematics, it was not clear how to relate θ to Ramsey's measurements. Some were skeptical that θ really did anything at all. Here, my friend Ed Witten appears again, once more with an important and precise experimental prediction. With several collaborators, he explained that, if QCD describes the strong interactions as they're observed, one can reliably calculate the neutron electric dipole moment for a given value of the number θ. They could show that θ had to be less than about a billionth if it was to account for these experiments. By now, the measurements are significantly better, and this pure number has to be less than one ten-billionth—and the limit will get even stronger over the next few years (or θ will be discovered).

This could just be a fact. But could there be a deeper explanation? Could there be some mechanism as a result of which θ was automatically extremely small? A possibility was put forward by Roberto Peccei and Helen Quinn in 1977. Both were then at Stanford. Their solution needed a new symmetry, which is now known as the Peccei-Quinn symmetry. Obtaining this requires structure beyond that of the Standard Model in its minimal form. In their original version, Peccei and Quinn had four new particles. What they didn't notice, however, was that one of their particles was necessarily quite light, lighter than any of the known elementary particles except possibly the neutrinos. This was recognized by Steve Weinberg and Frank Wilczek, who dubbed this particle the *axion*. The name was a play on a technical term, as well as on the name of a laundry detergent that enjoyed a brief advertising blitz in the late 1960s (I have a box of the detergent in my office, a gift from one of my former postdocs). Weinberg and Wilczek worked out the properties of this particle—its mass and the strength of its interactions with other particles—and it was soon realized that it was ruled out by a number of existing experiments.

Peccei continued a career working on phenomenon of the Standard Model at UCLA, where he eventually spent several years as the provost. Quinn went on to a career at SLAC, the particle accelerator at Stanford, where much of her work was devoted to the physics of the heavier quarks. She was a mentor to me in my time as a postdoc there. She also became a leader in the subject of physics education in grades K–12, retiring from SLAC a few years ago. Among her honors was membership in the National Academy of Sciences and award of the Franklin Medal. Peccei and Quinn shared the Sakurai Prize of the American Physical Society in 2013. Sadly, Peccei passed away in 2020.

Following the work of Weinberg and Wilczek, the axion idea lay dormant for a number of years. But eventually, various workers realized that the important feature of the Peccei and Quinn idea was not the details of their model but the light axion itself. As a member (in my case, a glorified version of the title postdoc) of the Institute for Advanced Study, I was working with colleagues Willy Fischler (then at the University of Pennsylvania and visiting the Institute) and Mark Srednicki (then a postdoc at Princeton) building some of the first supersymmetric models of particle physics, and we kept finding particles similar to the axion of Peccei and Quinn in our constructions. We quickly realized there was something more general going on. If the axion were lighter than in the original model, the interactions of the axion would be correspondingly weaker, and the axion would have escaped detection in experiments in accelerators. But other constraints on these particles, we learned, came from stars. Like neutrinos, if axions were produced in the cores of stars, most would simply pass through the star, carrying off some of the stellar energy with them. In fact, unless the interactions were *extremely* weak, axion emission would disrupt normal stellar processes. The strongest limits come from rather exotic stars—red giants, white dwarfs, and the last supernova visible with the naked eye from earth, SN1987a (for Supernova 1987a, also known as Shelton's supernova, for the person who first spotted it). When we published our paper, we learned that others had had similar ideas and had put forth other models for this "invisible axion."

My first response to our work was a certain element of pride, and also a bit of embarrassment. It seemed we had invented an excuse for a puzzle with no consequence. But Willy (now at the University of Texas, where he works on questions of quantum gravity, and moonlights as an EMT) started to pester me with questions about how this particle might behave in the early universe. This was new territory for me. I had never thought about issues in cosmology before. But as I was hiding out in the Physics Department at Brandeis while my wife attended a conference, I understood that the axion in the early universe would constitute, somewhat like the Higgs particle, a field pervading all space. Unlike the Higgs particle, this field would wobble around after the big bang, settling down only after a long time. This wobbling, Fischler and I realized, could be understood as a collection of a huge number of these light axion particles, all very nearly at rest. Willy pushed me to appreciate that, rather than being a disaster, if the axion mass were right, these axions could be the dark matter. It turned out that Willy and I weren't the only ones thinking about this question.

Pierre Sikivie and Larry Abbott, who had been postdocs with me at SLAC, and John Preskill, Mark Wise (both then at Harvard), and Frank Wilczek (whom we've encountered before) had come to the same realization. If these very weakly interacting particles were the dark matter, the question became: Could one detect them? Certainly not in particle accelerators—with their extremely weak interactions, the axions would be produced only very rarely, so rarely that it is unlikely any would be created in the collider—and *even* if some were produced, they would be virtually impossible to detect. But Pierre Sikivie appreciated that one might hope to take advantage of the great number of these particles in the universe—after all, we are assuming they constitute the dark matter and they are very light. In a very strong magnetic field, some of the dark matter axions will turn into photons. Sikivie's calculations showed that it would be feasible—though challenging—to build a detector with a large enough magnetic field and sufficient sensitivity to very small photon signals to detect the dark matter axions. Thus began what has been a long campaign that, in the last few years, has reached sensitivity to a range of interesting axion masses. The experiments have been driven by a group of skilled and determined physicists, most notably Karl van Bibber now at UC Berkeley and Leslie Rosenberg, now of the University of Washington. The experiment, involving about twenty-eight physicists from eight institutions, is known as ADMX, for Axion Dark Matter Experiment. Over the years, the collaboration has acquired or constructed progressively stronger magnets and detectors sensitive to progressively tinier photon signals. It's a remarkable instrument.

The photons one wants to detect, unlike the gamma rays we have spoken about for WIMP annihilation, have very long wavelengths, comparable to the wavelength of the radiation in your microwave oven, and very low energies. The strategy is to capture them in a cavity—basically a box with walls made of highly conductive material. The construction of these cavities is ingenious. So is the search strategy. The axion mass is not known, a priori, terribly well. It can vary over a factor of at least 100. Correspondingly, the energy of the expected microwave radiation is also not known precisely. So in the search, it is necessary to survey in very small steps over a large range of photon frequency. Think of dialing an old-fashioned radio with knobs, trying to locate a weak signal at a very precise setting on the dial. The experimenters have developed a protocol that allows for a systematic, painstaking search. If the simplest version of the axion dark matter idea is right, we should know within a few years.

But is it right? Even when we did our initial work, Fischler and I realized we were making some critical assumptions about the early history of the universe, which, while rather standard, have never been subject to experimental verification.

The most important of these is that the universe was once *extremely* hot, about 10^{25} degrees Kelvin. But we don't know from any actual observations that the universe was ever this hot. All we know from experiment is that the universe was once hot enough that nuclear reactions were common. This corresponds to a temperature of about 10^{10} degrees Kelvin, where the light elements were formed. In our original paper, we suggested an alternative to the standard cosmology, where the universe was never much hotter than this. In our picture, the axion would be much lighter. There would be correspondingly more of them in the universe (the density of the dark matter must be the same), but they would be much harder to detect. They could not be seen by the ADMX experiment. Over the years, better-motivated and more carefully developed models of such light axions have been studied. Most strikingly, it turns out that string theory predicts such light axions. Also, there have been some ideas for alternative strategies to detect these particles. Most recently, Peter Graham at Stanford and various theorist and experimentalist collaborators have put forward a proposal. Their idea involves exotic technologies and subtle physics. Prototype experiments are now in the planning and development stages. There is hope that they will open a window on this exciting possibility.

Searching for Dark Matter without Prejudice

All we know for sure about dark matter is how much of it there is, and that it hardly interacts with ordinary matter. It could be quite heavy—or extremely light. It could be a WIMP much heavier than the WIMPs that experiments like CDMS and Xenon1T might yet have seen. It could be something like an axion but much lighter, as we've suggested, which ADMX is not able to see.

Both the supersymmetric WIMP and the axion are striking in that they arise from answering other big questions about the laws of nature. But it is quite possible—even likely—that these proposals reflect the hubris of theorists as to their ability to guess solutions to big problems. The axion idea is, I have argued in various papers, the best we have to account for the strong CP problem. But it is not without warts, especially for axions in the range of masses probed by ADMX. The supersymmetry hypothesis has provided much of the motivation for searches for WIMP dark matter. But, as we've seen, the supersymmetry idea is in trouble at the Large Hadron Collider. The particles we would expect to see have not been seen. It's possible that we've been unlucky and that, in future running of the machine, we will yet find evidence for supersymmetry. It's also possible that our ideas are slightly, but not horribly, misguided, and we don't yet have quite enough energy to find these new particles. The WIMP might just be a somewhat heavier than expected version so that it would have escaped detection. But we should maintain a healthy skepticism about the ideas surrounding WIMP dark matter.

Stepping back to that question of hubris, one can justifiably ask: Have theoretical physicists gummed up the path to deeper knowledge with our prejudices about what principles should guide nature?

Many theorists and experimentalists have argued that we should discard any prejudices as to what the dark matter might be. There is much we don't know. Why should it be tied to any of the mysteries that currently trouble us? Perhaps we can catalog the possible forms of dark matter, and strategies to detect them. We can't realistically hope to cover every possible mass and interaction strength, but we can perhaps consider a broad swath of possibilities. This has led to a number of collaborations of theorists and experimentalists exploring the use of exotic technologies. To this theoretician, it seems a healthy, exciting development.

CHAPTER 11

THE DARK ENERGY

In Einstein's theory of gravity, energy bends space and time. In the space-time curved by a star or black hole, planets and light rays follow the path that is closest to a straight line. But these trajectories are shaped by the universe around them. The first tests of Einstein's theory looked for these effects, in most instances very small. It was only much later that objects like neutron stars and black holes were contemplated and discovered. Near these, space-time is dramatically altered. In the recent observation of gravitational waves from colliding black holes and neutron stars, we've seen the distortion of space-time itself.

Einstein contemplated more dramatic modifications of space-time when he thought about the universe as a whole. In his day, the energy that warped space was thought to be that of stars. The existence of nearby galaxies, much less faraway ones, was barely understood. In Einstein's model of the universe, space-time is warped so that the universe itself expands as time passes. As a result, everything moves away from everything else.

By now we know much more. Astronomers and astrophysicists have mapped out a universe full of galaxies, clusters of galaxies, and superclusters. In between, there are large quantities of gas, mostly hydrogen. And as we've seen, most of the mass of the universe is dark matter.

Einstein's Greatest Blunder Revisited

But there is an additional possible form of energy, which, from the beginning, loomed as a monstrous possibility—monstrous both in its quantity, and its possible effects on space-time. This would be a certain amount of energy at every point in space and time. This is something allowed by Einstein's theory—it is what he called the cosmological constant, the term he added to his equations in his initial hopes of avoiding the expansion of space and time. We've seen that, with Hubble's discovery of the expansion of the universe, Einstein attempted to banish this cosmological constant, declaring his earlier attempt to include it as the greatest mistake of his career.

But why did he think it such a mistake? As best as I can tell, he thought they made his equations look ugly, and, at least for a while, it wasn't clear that it was needed to match the data on the growth of the universe. But nature's laws aren't obligated to look beautiful according to someone's aesthetic, even Einstein's. If Einstein hadn't been so hostile to quantum mechanics, he might have been puzzled not that a cosmological constant wasn't there but why it wasn't huge. So huge that it would curl the universe into a little ball, perhaps a few kilometers across, or possibly much, much smaller.

What is this confrontation between quantum mechanics, which rules the world of the very small, and scales the size of the universe? The problem is the uncertainty principle. Just as the uncertainty principle says that one can't know the position of a particle precisely without sacrificing knowledge of its speed or energy, so for every way light can vibrate, one can't know there's no vibration at all. There is always a little bit of energy for every possible vibration—every possible frequency (kilohertz on your radio). The number of possible frequencies is infinite, as far as we know. So everywhere in space, there should be an infinite amount of energy. Just as strange, this energy comes with a pressure that is *negative*. That may not strike you as weird. For most of us (I confess, even for lots of my colleagues and students, who should know better), pressure is not a concept we understand too deeply. Suffice it to say that in a balloon, for example, it's the pressure of the air inside, pushing against the wall of the balloon, that keeps the balloon from collapsing. The same is true of the tires on your car. With negative pressure, it would be as if the gas was sucking the walls in.

You might—quite reasonably—ask, Why, given this crazy result, don't we throw quantum mechanics away? This is a question that occurs to every physics student when they first encounter this feature. But what their teachers and textbooks reassure them is that this infinite energy has no consequence. The teachers (me included) get away with this because, before adding general relativity to the mix, photons, electrons, protons, and everything else move through this soup of energy totally undisturbed.

For teachers of my generation and older, there was another excuse for ignoring this problem. General relativity long had a rather shaky reputation, and certainly combining it with quantum mechanics was a shady business. So, if we were aware of this problem at all, we just closed our eyes to it. If we needed to, we allowed ourselves to sleep thinking that, perhaps, there's just some other energy, coming from somewhere, that just cancels this out.

I was forced to confront this problem rather early on by Leonard Susskind. He had already shaken my worldview by introducing me to the hierarchy problem. A bit later, he hit me with the even more extreme problem of fine-tuning. In Aspen, Colorado, in a spectacular mountain setting and adjacent to the homes of celebrities and the very wealthy, there is a small physics institute where theorists gather in the summer, usually for periods of about three weeks. It's quite beautiful. During the week, they work; on the weekends, they hike or bike in the nearby mountains. One summer, as a postdoc, I went for a visit. My friend Willy Fischler was there, as was Lenny. I remember Lenny asking me why the universe wasn't rolled up into this tiny ball (he was willing to contemplate that it was the size of the moon). My innocence was lost. My all-knowing being, dialing the knobs to create a suitable universe, was forced to adjust not to the 32-decimal place but to the hundredth. The hierarchy problem was starting to look minor.

Later, when Willy and I, along with our colleague Mark

Srednicki, were hopeful we had solved the hierarchy problem with supersymmetry, we thought we'd better solve this other problem as well. In fact, supersymmetry was rather neat from this vantage point. If supersymmetry weren't broken, the cosmological constant might be zero. The contributions of the different types of particles would cancel out. But supersymmetry can't be exact, and we realized we would still need at least a cancellation of 60 decimal points. We racked our brains trying to solve this puzzle, and imagined the answer was within reach.

But apart from looking for a solution, there was another aspect of this puzzle. We knew, ultimately, the number had to be extremely small. The universe, after all, is quite big. So maybe it was just zero. If astronomers were to measure such a thing, it would have to be just such that it is only barely important when the universe is as old as it is now. Why now?

Nature, however, was about to have the last laugh.

How Could the Universe Be Younger than the Oldest Stars?

Like many, I live in a household with two working spouses (and once, young children). We've always needed to live near my wife's work, so I have always had long commutes. I can't stand to drive alone and, in any case, am racked with guilt when I do, so, as I've mentioned, I have generally used public transportation or carpooled. For the many of you who suffer similarly, I recommend this transportation plan. It has done much to improve my emotional health, and has meant that the time commuting was not simply lost. Over the years, it significantly reduced my carbon footprint as well. One of my car pool partners through the years has been George Blumenthal. George is a distinguished astronomer, who played an important role in the development of the theory of dark matter. He later became chancellor of my university. But in the 1990s, he had become my neighbor, as *his* spouse, a law professor, Kelly Weisberg, taught in San Francisco and they were trying to balance their commutes. George and Kelly's children were the same ages as my older two, and our rides to work, when not consumed with conversations about our kids or campus politics, were occupied with discussion of a curious puzzle. In Einstein's theory, assuming that the matter and energy we see (including the dark matter) is all there is, then from the measurement of the expansion rate of the universe today, one can determine the age of the universe, i.e., the time that has elapsed since the big bang. In the mid-1990s, measurements of the Hubble constant gave the age of the universe at about 9 billion years. But this posed a serious problem. There were objects in the universe called globular clusters that were known to be older than that, about 10 billion years old. Needless to say, this was quite an embarrassment.

George told me about one solution, to which I rebelled. It turns out that if there is a cosmological constant, the expansion of the universe is speeding up (astronomers speak of acceleration in this context). This means it was slower in the past than we would have otherwise thought, and the universe is older than it would be without the acceleration. A number of workers pointed out that this could provide an explanation of the age discrepancy. To me and many others, a cosmological constant of just the right size to do this seemed a big price to pay. But there was another motivation to consider a cosmological constant. This came from work on the formation of galaxies, and it is something I was hearing a great deal about from my Santa Cruz colleague Joel Primack. He, along with various collaborators, was trying to understand how galaxies form. They assumed, when they began their work, that the energy of the universe through most of its history is dominated by matter—ordinary matter and dark matter. To study this problem required what were then state-of-the-art computer simulations. What they found wasn't far off from what is observed for typical galaxies, but it wasn't exactly right either. If, instead, they modified their models, supposing there was a cosmological constant and that this contributed a significant fraction of the energy today, the results were better.

To appreciate why this seemed so strange to me, one can ask when, in the history of the universe, would a cosmological constant of this size be important. The answer is that for the first few hundred million years, it was of almost no consequence. When the universe was a billion years old and stars were beginning to form, its effects were barely perceptible. For some reason, if my colleagues were right, the cosmological constant is such that it becomes important only "now," at our current epoch in the history of the universe. Once again, I was skeptical. It seemed to me much more likely that one had misunderstood some aspects of globular clusters or galaxy formation than that one had this crazy number just right to be important in the universe now and in the future—but not before.

My prejudices were just prejudices, and they turned out to be completely wrong. Fortunately, astronomers and astrophysicists were aware from the early days of general relativity of the possibility of a cosmological constant, and they looked for it. Because the universe is not curled up into a little ball, they knew it could not be too large. By the early 1990s, they had no positive evidence for it; they could only say that it must be smaller than some number, such that one could barely hope to observe it.

Steven Weinberg here enters our story again. Weinberg was a scientist of remarkably broad interests. In addition to putting forward the critical elements of the Standard Model, he became an expert cosmologist. To learn the subject, in the early 1970s, he taught a course and wrote a comprehensive textbook, *Gravitation and Cosmology.* He wrote numerous papers as well, some of which would serve as the starting point for calculations of the amount of dark matter. In recent years, to bring his knowledge up-to-date, he gave another course and wrote *another* textbook, *Cosmology.*

Weinberg was very interested in the question of the cosmological constant, and the problem we have discussed of why it is so small. In 1989 he gave a set of public lectures at Harvard and wrote a review article on the subject. He took crude arguments and made them sharp. Rather than simply say, "The cosmological constant can't be too big or the universe would be very small," he pointed out that if the cosmological constant was too big, stars and galaxies would not have formed. More precisely, galaxies start to form about a billion years after the big bang. If the universe was expanding too fast at the time, the material that would otherwise form stars would be blasted apart by the explosive expansion. So, even without direct observations, he could say how large the cosmological constant could possibly be. Turning this around and saying it's hard to understand why the cosmological constant is small, he argued that the cosmological constant might not be much smaller than a certain value. Observing this value would require studying the motion of stars and galaxies billions of light-years away from us.

Two groups of observational astrophysicists, one led by Saul Perlmutter at UC Berkeley and one by Adam Riess of Johns Hopkins University and Brian Schmidt of Australian National University, devised a strategy to measure the cosmological constant. They realized that the question could be answered by studying a particular type of violent event, *supernova explosions*. Supernovae, the explosive ends of the lives of some stars, are among the most dramatic events that occur in the cosmos. Astronomers understand quite well the birth and death of stars, so these events can be tools to make measurements about the universe. Stars are born in regions where there is more than the average amount of material—typically hydrogen gas but with some mix of heavier elements. The particles start to grow closer due to their gravitational pull. Once enough of them are packed tightly together, they heat up. Eventually, there is enough heating that nuclear reactions start to take place, and the star becomes hot, giving off light, heat, and other radiation. From this point in time, the life of a star is a competition between the thermonuclear reactions occurring deep in their interiors, tending to blow them apart, and their enormous gravity due to their large mass, which wants to pull all the stellar contents into the center. The balance between these forces determines the size and temperature of the star. Eventually, the nuclear fuel runs out, and the star collapses due to gravity. (The material doesn't actually disappear, but there is finally not enough hydrogen to maintain the basic reaction.) At this point, different things happen depending on the size (mass) of the star. Our sun will go through different phases, exhausting most of its nuclear fuel in about 5 billion years and becoming an enormous, hot red giant for a comparatively brief period before ending its life as a cold remnant known as a white dwarf.

Other stars, with larger masses, at some point exhaust their nuclear fuel and collapse; all the material near the surface of the star rushes toward the center. Now, once the material is packed tightly together, the system heats up again, rapidly, and a massive explosion results. This is known as a Type II supernova. Much of the material is blown out into space, where it can eventually become the stuff of other stars, providing, in particular, heavier elements like carbon and iron. Some of the material emerges as neutrinos or light. Much of the remainder is left behind as a neutron star, or sometimes a black hole. Supernovae that can be seen from earth with the naked eye are rare. The last was observed in the Southern Hemisphere in 1987. A second type of supernova explosion arises in binary systems, when a white dwarf collects material from another star and reignites. These are known as Type I supernovae.

What is remarkable about the Type I supernovae, and made them so useful for Riess, Schmidt, and Perlmutter, is that they are what astronomers call *standard candles*. They are all very nearly the same, and as a result, the light they emit is a distinguishing characteristic. Most important, the spectrum of radiation of the emitted light is known. The wavelength of the light we observe on earth is shifted, depending on where and when it was emitted, according to Einstein's redshift rules. Perlmutter, Riess, Schmidt, and the members of their collaborations realized that if one observes Type I supernovae across the sky and measures the radiation of different frequencies, then one can make a map of the gravitational field of the universe. This information can be used to determine the rate of expansion of the universe at different times.

Their result yielded a cosmological constant of just the size required to account for the age of the universe and the formation of structure, and was roughly what Steven Weinberg's argument had predicted. The dark energy, in fact, currently makes up the bulk of the energy of the universe, about 70 percent. Perlmutter, Riess, and Schmidt were awarded the Nobel Prize for their work in 2011.

In subsequent years, further observations of supernovae, as well as an array of other observations, have confirmed the discovery. I have spoken of the result as a cosmological constant, though physicists and astronomers tend to call it the *dark energy*, since it could conceivably be something else. Despite my chapter title, I want to continue to call it a cosmological constant for two reasons. First, there is, by now, good supporting evidence that not only is the energy density what one expects, constant over the universe in space and time, but so is the pressure. Second, as we'll discuss later, the cosmological constant is crazy enough. Anything else is far crazier.

We are at a momentous stage in the history of the universe. Since seconds after the big bang to the present epoch, the expansion of the universe has been slowing down. When the universe was 150 years old, it was about 10,000 times larger than it was at 3 minutes. In the subsequent 13 billion years, it's grown only by a similar factor. All that is about to change. In the next 25 billion years, the universe will grow 10 times larger. In the 25 billion years after that, it will grow by another factor of 10. In the next 25 billion years, another factor of 10, and so on. In other words, in 100 billion years, it will be 10 billion times larger than it is now. In a trillion years, it will be 10^{100} times larger than it is now—a *googol* times as large. The galaxies that are a million light-years away from us now will be a million googol light-years away, totally beyond anything that could be seen with instruments. On such time scales, the universe will be a bleak place.

Time to take a step back from that cold distant future.

AND STEPPING INTO
THE UNSTABLE

CHAPTER 12

AT THE BEGINNING
OF EVERYTHING

We know that the universe was once far smaller than what we see today. We can reliably trace its history back 13 billion years, when it was twenty orders of magnitude—twenty powers of 10—smaller than it is now. We refer to this first instant as the big bang. But did the universe really shrink to a point? How far back can we look? To the latter question, astrophysicists have an answer. Right after the big bang, before the baryons and the dark matter were created, something very dramatic happened. We have very extensive evidence for this event, but it remains in many ways quite mysterious. It is what is known as the epoch of *inflation*. This is quite possibly the earliest time for which we can hope to acquire direct evidence.

In the last chapter, we learned that the universe is entering an era of exponential growth. Every 10 billion years or so, it will grow by another factor of 10. But what's perhaps even stranger is that this has happened before. A tiny fraction of a second after the big bang—maybe 10^{-33} seconds after the big bang—the universe went through a period when it grew by a factor of 10 every 10^{-33} seconds (that's about a trillionth of a trillionth of a billionth of a second), and it did that long enough that the universe grew by a factor of 10^{100}—a googol—give or take a few factors of 10. This is like a bacterium growing to be the size of the universe, and then doing it three more times, all in a tiny fraction of a second. Why would cosmologists think this? And how could we possibly know? This is a case where apparently eccentric ideas can become the subject of precise study and experiment. In the late 1970s, Alan Guth was a struggling young postdoc. He was focused on a set of problems in big bang cosmology, questions far from the mainstream of theoretical physics at the time. I remember well his describing these when he visited SLAC one summer when I was a postdoc there.

Three things concerned Alan. The first problem was a sharp version of the unease I felt when I first learned of Einstein's starting point for cosmology: The universe is the same everywhere and in every direction, the *cosmological principle*. As we've seen, this statement holds true to a remarkable degree. We know that the temperature of the microwave background, for example, is the same in all directions to a part in 10,000. If in one direction it's 2.7001 degrees Kelvin, in another it's 2.7002. At the time that Guth was worrying about this question, things were not known nearly so precisely, but it was clear that the temperature was at least roughly the same in whatever direction one looked.

At some level, this is not so surprising. If you pump hot air into a container containing cold air, and seal up the container, at first the air in the different parts of the container will have different temperatures, but after a little while, the temperature of the gas will be the same everywhere. The issue, for the temperature of the universe, is in the "little while." For the air in the container, temperature is a measure of the average energy, or speed, of the molecules. At first, when some of the air is hotter than the rest, the molecules in that neighborhood are faster than those in the other bits of the container. But the fast molecules collide with slow molecules, giving them some of their energy until, after a while, they all have similar energy—and the same temperature. The time it takes for this to happen is controlled by how often the molecules collide. If we take the temperature at various points in the container very quickly after we pump in the hot air, we will get different results.

The fastest that different parts of the container could come to the same temperature is the time it would take light to connect them. This is the issue for the early universe. When we look at the CMBR (cosmic microwave background radiation), we are looking back at something like 10^{12} regions that couldn't have talked to each other, even at the speed of light, when the universe was 100,000 years old, but they all have almost exactly the same temperature. Was the universe just created this way?

A second puzzle Guth considered has to do with the curvature of space. We're now familiar with the idea that space-time is warped in the history of the universe, but it is also possible that, at any given time, space itself is bent. This can be understood by analogy. If we're an ant living on the surface of a ball, our world is essentially two-dimensional, but a curved two-dimensional world. If we travel far enough on a great circle, we come back to where we started, for example. It turns out that in Einstein's cosmology, something similar is possible, just with one extra dimension. If we look around us, it's clear that the universe is pretty flat. It turns out to be hard to understand this. At very early times, things have to be just right to avoid a large curvature of space.

Guth was also bothered by a third puzzle, known as the *monopole problem.*

This part of the story of inflation gives me a chance to sneak in another question. We've spoken about the difference between electric and magnetic forces and gravity. The electric forces in an atom are far, far larger than the gravitational forces. But in nature, most atoms and collections of atoms are electrically neutral, so at large distances, these forces cancel out, and gravity wins for large objects—planets, stars, and even much smaller things. Critical to this is that the electric charge of the electron is exactly opposite to the charge of the proton. We should be careful when we say "exactly" to ask, experimentally, how precisely we know this to be true. Without sitting down to do a dedicated experiment, we can make a rough guess. If, say, the charge of the electron were slightly smaller in size than the charge of the proton, the sun and the planets would all carry a positive charge, equal to the sum of the positive charges of all their atoms. This would lead to a repulsive force between, say, the sun and the earth. This force would win out over gravity, pushing the sun and the earth away from each other if the excess charge was roughly a part in 10^{19}. More detailed considerations give a much stronger result. So it looks like these charges are exactly equal and opposite to one another.

As so often in thinking about laws of nature, we can ask: Is this just the way it is, or can we offer an explanation? Here, Paul Dirac enters our story again. For most physics students, including myself, electricity is a lot easier to think about than magnetism. This is because, for electricity, all one has to do, basically, is follow the electric charge. Magnetism, on the other hand, is harder to understand. It arises from electric currents and from the spins of particles. The equations are more complicated. It would be simpler to think about magnetic charges, if such things existed. But no one has ever found one, and Maxwell's equations refer only to electric, not magnetic charges. Such isolated magnetic charges are called *magnetic monopoles.*

In 1931 the same Paul Dirac who proposed the existence of antimatter asked what would happen if one added monopoles to Maxwell's theory. If it had been me, I would have done the obvious thing and just stuck a magnetic charge in Maxwell's equations. Like Einstein developing a relativistic theory of gravity, however, Dirac thought about this problem in a much deeper and more subtle way. Dirac realized that the existence of a magnetic monopole—even one—somewhere in the universe would pose a challenge for quantum mechanics. He considered the Schrodinger equation in a world with charged objects like electrons and a single monopole. He realized that the Schrodinger framework would not allow both types of particle without some mathematical sleight of hand. In fact, his remarkable paper begins with a manifesto: "The steady progress of physics requires for its theoretical formulation a mathematics that gets continually more advanced."

Dirac then argued that quantum mechanics doesn't work if

there's even a single monopole somewhere, unless all electric charges are related in such a way that the proton and electron charges exactly cancel. (Physicists say that electric charge is quantized; all charges are rational multiples of a basic unit of charge.) This was the first explanation ever offered for the remarkable fact of charge quantization, and since that time, experimenters have been open to the possibility that monopoles might exist. Various searches have occasionally turned up candidates, but to date they have turned out to be ordinary phenomena faking a signal. On the more theoretical side, there have been significant developments. In Dirac's approach, monopoles are introduced in an ad hoc way. One doesn't know their charges, their masses, or any other properties a priori. Grand unified theories and string theory are homes to monopoles—they arise automatically, as part of the theories—and they obey Dirac's rule. Most of the time, these particles are extremely heavy. They're also stable. So there is a good chance that such things exist. While they're almost certainly too heavy to produce in accelerators, there may be places and times in the cosmos where monopoles have been created.

There are very strong limits on the number of monopoles. One arises from the existence of magnetic field that pervades our own galaxy. Monopoles, if they existed, would tend to cancel out this field. A second comes because we know pretty well how much energy is in the universe. In theories like string theory and grand unified theories, the monopoles are extremely heavy, typically 10^{16} times the mass of the proton or even more (again, this is about 10,000 trillion). Since we know how much energy carried by matter is in the universe, we can say that there can't be more than about 10^{-16} as many of these objects as there are atoms. On average, then, your chance of finding one monopole in a typical cubic kilometer of the universe would be about one in 100 million.

But now we encounter the big problem. If the universe was extremely hot in the past, at that time, it would have produced monopoles and antimonopoles. A large fraction of these would survive as the universe cools. Today they should exist in numbers comparable to the number of photons, and given their huge masses, this is unacceptable. Even if the universe was never quite so hot, there are general arguments that lots of monopoles should be produced. So, either monopoles play no role in nature or something has to give with our understanding of the cosmology of the extremely early universe.

Inflation

In 1981, Alan Guth put forward an idea that solves the monopole problem, and also the problems of homogeneity, isotropy, and flatness. He called his proposal inflation, by analogy with inflation in the economy. The 1970s were a time of rather high inflation in the United States, with worries that the country might experience "runaway inflation." In Germany in 1923, an extreme case, prices rose at 41 percent per day. At that rate, prices increase by more than a factor of 10 every week (fortunately in the United States, things have never been like this, except perhaps in the Confederate States toward the end of the Civil War). Then if a loaf of your favorite bread costs $5 this week, it will cost $50 the following week, and $500 the week after that, and so on. After a year, your favorite loaf of bread will cost an unimaginable sum, 10^{52}. This is a number of dollars comparable to the number of atoms in the sun! Of course, things could never get this far. The money would simply run out. In practice, there would likely be a change of government, or worse.

Guth speculated that the universe might grow in size in a similar manner, for a short period. If the universe grew by about 60 orders of magnitude or more, this would solve all the problems we've listed. The fact that the temperature is almost the same everywhere would be automatic. All of the universe we can presently see was originally packed into an incredibly tiny volume, so it is natural that everything we see is at the same temperature. Even if the universe started out as something like a sphere, it would have completely flattened out. And whatever monopoles there might have been were diluted. At most, there would be one monopole in the entire universe.

Guth put forward a model for this phenomenon inspired by ideas of grand unification. In his picture, the universe started off extremely hot, and at some point, inflation turned on. Guth quickly realized that the model he put forward didn't do what he hoped. Inflation never stopped. But a new proposal was soon put forward by Andrei Linde (then of the Landau Institute in the Soviet Union, now at Stanford) and Paul Steinhardt (then at the University of Pennsylvania, now at Princeton) and his student Andy Albrecht (now at UC Davis). In their model and in many variants that have been studied since, inflation switched on and switched off. The model had several working parts, including a new field, somewhat analogous to the Higgs field, called the *inflaton*. While the theory had its warts, it worked. The whole story was dubbed the "New Inflationary Universe."

While the inflation hypothesis would resolve the problems of the big bang theory, at first it seemed to leave no distinct calling card. This changed with the work of Guth and So-Young Pi and of James Bardeen, Paul Steinhardt, and Michael Turner on the quantum version of the inflation theory. When combined with quantum mechanics, inflation turned out to provide an explanation for one of the great, long-standing puzzles of physics and cosmology: How did structure—galaxies, stars—arise? And with this explanation came a prediction for the cosmic microwave radiation.

Perfect homogeneity and isotropy isn't good, as we've said. We'd like some irregularity to seed the formation of the structure we see. But it was soon realized that inflation, while making everything almost homogeneous, necessarily gave rise to a universe that was not perfectly smooth. The origin of the *inhomogeneity* is quantum mechanics. In quantum mechanics, the uncertainty principle is crucial. In its simplest form, the principle says that one can't simultaneously know the position and momentum of a particle with arbitrary precision. And the principle has much broader applicability. In the context of theories of inflation, it says that one can't know precisely the value of the inflation field and the rate at which the field is changing at any given point in space-time. Yet it is this combination that determines how much energy there is at any point in space. So while the theory predicts the average amount of energy in the universe, it also predicts that there will be small, random variations—quantum probabilities again—in the amount of energy from place to place. As it happened, without knowing too many details about the underlying theory, these two groups were able to calculate the predictions of inflation for the variations of the energy density—and with this the variations of the temperature.

New inflation turns out to be not one model but a broad class of models. Each makes a specific prediction for the temperature variation, but no one is so compelling that one can say: This is the number the theory predicts. What we do know is that stars and galaxies started to form about a billion years after the big bang, and from this, the variations in the temperature should be about a part in 10^{-5} or perhaps slightly larger. Now the question was: Could one find them? Early studies, as we have seen, saw no variation. The first observations came in 1993 from a satellite called COBE (for Cosmic Background Explorer). COBE observed tiny variations in the temperature—about a part in 100,000, just large enough to account for the structure we see in the universe. The COBE principal investigators, John Mather and George Smoot, were awarded the Nobel Prize for this work.

Often in science, the first observations of a phenomenon are just on the edge, barely significant enough to make a claim of discovery, and subject to skeptical criticism. But if a phenomenon is real, it is frequently followed by improved observations, with better instruments and techniques, which permit convincing confirmation and detailed study. For the CMBR, this has been true in spades. A sequence of satellite and ground observations have studied the CMBR in great detail. Thanks to satellites such as the Wilkinson Microwave Anisotropy Probe (launched by NASA in 2001) and Planck (launched by the European Space Agency in 2009), we have detailed maps of the temperature across the sky. This data can be unraveled to reconstruct a great deal of information about the early universe. It provides, for example, a precise measure of the energy budget—the dark energy, the dark matter, and the ordinary matter.

What's the Data Telling Us?
What Would We Like to Know?

This wealth of data confirms, first, that the big bang really happened, and second, that the universe once had a temperature of about 10,000 degrees Kelvin. At this time, the universe was, as Einstein assumed in his first attempt at a cosmology, homogeneous and isotropic to a high degree. And much more than that, we now know with great confidence that the structures we see around us—stars, galaxies, clusters of galaxies—grew from small violations of this isotropy and homogeneity. From this, we have good reason to believe that inflation happened.

Particle physicists would like more. They would like to understand precisely what laws of nature—what fields, what equations for these fields—account for inflation, and perhaps from this have a glimpse of what came before. Again, while inflation is very successful, it is not really a theory but a class of theories. One thing we don't know is exactly when inflation happened, or when or how it ended. We're pretty sure it must have ended before the universe was a few minutes old. But the most dramatic period, the period of the most rapid inflation, probably occurred much earlier. Inflation is characterized by the amount of energy at the time it happens. This, in turn, translates into the doubling time for the size of the universe. For example, if the energy density was that characteristic of grand unified theories or string theories, this doubling time is extremely, unimaginably short—10^{-37} seconds, or 0.000000000 00000000000000000000001 seconds. In other words, our satellites would be providing us with a picture of the universe at an infinitesimal time after the big bang. How could we know? One possibility is that we will find some compelling theory that explains a great deal of data and makes a sharp prediction for these phenomena. There are many proposals, but as yet, none is particularly convincing.

There is, however, the hope that experiment will settle some of these questions, especially the energy scale of inflation. Provided this scale is not too low, maybe a factor of 100 below the scale of grand unification, it should be possible to observe gravitational waves from inflation. These waves would be produced by quantum effects, much like the tiny inhomogeneities in the energy which account for structure. But gravity—and gravitational waves—are generated by energy, so the less energy, the fewer gravitational waves. In 2014, with great fanfare, an experiment at the south pole, called BICEP2, announced the observation of gravitational waves from the early universe. Much scrutiny and criticism followed. It turned out that dust (again, astronomers call just about any type of particle—hydrogen and other atoms—dust) could account for the BICEP2 observations. The claim was promptly withdrawn, and improved searches and analyses that can account for the dust are in progress. Within a few years, we might know when inflation happened. While not compatible with my current theoretical prejudices, it is quite possible that none of these experiments will see a gravitational-wave signal, and at best we will be able to say that inflation occurred after about 10^{-36} seconds. But imagine if we can reliably say that humankind has glimpsed the universe when it was 10^{-36} seconds old!

When inflation ended, the universe was very hot, with a temperature almost exactly the same everywhere. This raises a puzzle: How did all the inhomogeneity we presently see— galaxies, stars, planets—emerge from this uniform soup? The answer has to lie in the tiny variations in the energy predicted by Guth and Pi, which gave rise to the tiny variations in the temperature. Guth and Pi realized that these small variations would grow, becoming large enough, after a while, that regions of higher density would collapse and start to form galaxies and stars. They worked out the theory before the measurements of the temperature variation, and *predicted* this 10^{-5} number. (There was some uncertainty because they did not have a detailed model for how galaxies would form from this variation.)

Remarkably, these small fluctuations in the density are the result of quantum mechanical fluctuations—essentially a realization of the uncertainty principle.

The details of how galaxies form are complicated. The dark matter is a critical component, as first explained by my Santa Cruz colleagues George Blumenthal, Sandra Faber, and Joel Primack in 1984. In their picture, the dark matter clumps first, drawing in ordinary matter—mainly hydrogen gas, which starts to collapse, forming stars and galaxies. By now, the study of galaxy formation, aided by powerful computers and ingenious algorithms, has provided us with a convincing and detailed picture of how the structure we see in the universe formed from these small quantum mechanical seeds right after inflation. We've spoken of quantum mechanics and the world of atoms and things smaller. Now we see that quantum mechanics is writ large on the sky!

The Very Early Universe: Where Do We Stand?

From the observations of the cosmic microwave background and the formation of structure in the universe, we can be pretty sure that the phenomenon Guth called inflation happened. It is tantalizing that we have some knowledge of phenomena at such early times, but we don't have more than a gross picture. Inflation is almost certainly pointing us to new laws of nature and possibly new overarching principles. A discovery of gravitational waves from inflation would give us important clues as to what is going on. But even then, we will probably need some theoretical breakthrough, perhaps from string theory, perhaps from some other direction, to fully sort out what is going on.

Other big questions loom here. Inflation is like a curtain hiding what happened when the universe was still younger. We still don't know whether, at times before inflation, space-time shrinks to a point or is replaced by something very different. Here we enter realms of wild, and largely ungrounded, speculation. Follow me.

CHAPTER 13

CAN WE GET TO A FINAL THEORY WITHOUT GETTING UP FROM OUR CHAIRS?

Most of us are taught in school that the Greeks and other ancient societies believed that the laws of nature could be inferred by pure thought, without experiments or careful observations. We learn that after Galileo, most humans have thought of the laws of nature as something revealed by experiment and observation. Indeed, Newton's laws followed a century after Galileo's work from the struggle to understand the observed motion of everyday objects and of the planets. The laws of electricity and magnetism were the result of more than a century's careful study of electric charges and currents. The important numbers in nature—the masses of atoms, the strength of Newton's gravity, the electric charge of the electron—are things that have been measured in laboratories, not determined by some abstract reasoning. But perhaps it doesn't have to be this way. Perhaps the laws of nature, the equations, the constants in books are inevitable, the output of some set of grand principles.

As a student, my teachers warned me that big ideas in science have rarely been uncovered through grandiose theorizing. Einstein's general relativity, they acknowledged, is an exception. Experiments and observations did not drive his quest. The theory, instead, follows from a grand principle: The laws of nature should look the same to any observer anywhere, moving in whatever way they may be. From this simple starting point, an awesome structure emerges. But, with fingers wagging, they said—don't imagine going there. First, you're not nearly as smart as Einstein. Second, Einstein wasted much of the last half of his life trying, and failing, to replicate his earlier triumph.

I tend to respect authority. But the temptation to pursue such an ambitious agenda has sometimes proven irresistible to many of my colleagues and occasionally to me. While perhaps not quite as dramatic as the general theory, some other examples of theoretical leaps driven by pure thought have given me pause. One is in the work of Maxwell. While the development of the laws of electricity and magnetism were closely tied to experiments, Maxwell took a step guided at least in part by grand principle. Among the laws of nature we've discussed is that electric charge is conserved; it is not created or destroyed. I am not sure how good the experimental evidence was for this rule in Maxwell's day, but Maxwell thought it should be true. He took what was known of the laws of electricity and magnetism at the time, and saw that they violated this principle. He might have argued that electric charge is not always conserved, and that one should test this rule experimentally, but instead, he modified the laws so that electric charge *was* preserved. From this starting point, he predicted the existence of electromagnetic waves.

Certainly the quantum theory was not anticipated by any theoretical reasoning—and it is hard to imagine how it might have been. The experimental exploration of the world of atoms indicated that not all was well with classical physics. But I have always been awed that the early quantum mechanicians sorted out what was going on. How in the world did anyone think that one should introduce probability into the whole structure? In some sense, the critical step was one of pure thought. It came through the work of Max Born, a German theorist. Born was thinking about experiments, but not so much particular experiments as how the theory could cope with a broad class that, like Rutherford's, employed some sort of projectile, perhaps an electron, colliding with a target, say an atom or an atomic nucleus. In trying to figure out what Schrodinger's equation said about such processes, he hit upon the idea that Schrodinger's wave function is related to probabilities of different outcomes, and even stranger, that one had to take the square of the thing to actually determine the probability. One could conceive of arriving at this result through the experimental facts of radioactive decay, but here, this most critical (and puzzling) feature of the quantum theory came from purely theoretical reasoning. Bohr, Heisenberg, and Dirac had closely related ideas and quickly adopted this viewpoint, but Schrodinger and De Broglie could never bring themselves to accept it. Einstein, as we've seen, was never comfortable with the role of probability, and after his early work on the photon and his work on the behavior of large aggregates of quantum particles with Satyendra Nath Bose, he did not make significant contributions to quantum mechanics.

Still, Einstein's triumph with the general theory is the most spectacular example of such a success, and Einstein believed—or at least hoped—that all the laws of nature should emerge in a similar way. In the latter half of his life, he embarked on a quest for what he called a unified theory, a set of principles from which all the then-known laws would follow. This indeed was a slippery slope. Einstein wasted years of his life on what turned out to be a fruitless quest, as have many other, lesser scientific minds.

Despite this history, many theorists have been seduced by the tantalizing prospect that string theory could fulfill Einstein's dream of a unified theory (sometimes called a Theory of Everything). If so, the discovery of string theory is truly serendipitous. Unlike Einstein's general relativity, it's not, in its current formulation, held together by a simple, compelling principle. Instead, it's a sort of Rube Goldberg contraption, cobbled together from parts collected from a scrap heap of theoretical ideas. Yet the output is somehow striking and beautiful.

The clash between quantum theory and general relativity takes various forms. Some seem technical but nevertheless hard to surmount. One, raised by the late Stephen Hawking, is more conceptual: The whole notion of probability, so central to quantum theory, seems to break down in the presence of a black hole. String theory resolves all of these questions. It also seems to be able to explain the features of the Standard Model, the history of the universe, the dark matter and dark energy, and much more. But it's not clear it's right, or that it even makes definite predictions at all.

While some take all this as a sign that humanity has stumbled on a Final Theory or Theory of Everything, which might account for all the known laws of nature and predict new ones, others have argued that this is just one more dead end. Many readers will know that string theory has been a lightning rod for criticism. In this chapter, we'll understand why, on the one hand, the subject is so seductive, and on the other, its critics may have a point.

In current theories, like the Standard Model, the basic objects—the electron, quarks, neutrinos, photons—are just points in space. This sounds like an idealization, but amazingly, both in our theories and in experiments, there is no indication that they need to be anything more complicated. The study of elementary particles with accelerators contrasts sharply with the worlds revealed by early light microscopes. The very first microscopes uncovered all sorts of intricate structure when drops of water and samples of living tissue were examined on small scales: microorganisms, cells, and much more. Modern particle accelerators act like microscopes with very high magnification. But there's no evidence of any analogous structure when electrons, quarks, and the other particles we call elementary are resolved at scales of a millionth of a trillionth of an inch.

Still, we might think this picture of the basic objects in nature is somehow naive. We could imagine replacing points with something very complex, like cells of living organisms, but let's instead try something just slightly more complicated than a point: Suppose the basic entities in nature are not points

253

but lines, either line segments with ends or closed lines without ends, like rubber bands. We'll call these entities *strings*. Such strings can vibrate in different ways. If we hypothesize that the strings must obey the rules of special relativity and quantum mechanics, amazing and bizarre things happen. Each of the modes of vibration corresponds to a different type of particle. Among them are gravitons, the particles that carry the force in Einstein's theory of gravity. There are also quarks, gluons, electrons, photons, and more. The particles interact according to the rules of the Standard Model and general relativity. The theory has no problem with quantum mechanics. Getting the details right—the exact set of quarks and leptons we see, the Higgs mass, the observed value of the cosmological constant—is the challenge.

The fact that this seemingly dumb starting point yields two of the towering achievements of twentieth-century physics, the building blocks of the Standard Model and of general relativity, is what gives some theorists pause.

So we can swallow hard, accept that the basic entities *might* be lines rather than points, but why should we be so impressed? To understand what gets some physicists so excited, we need to understand why quantum mechanics and Einstein's theory of gravitation seem to be in conflict, and whether the stringy view of fundamental physics might reconcile them.

Given Einstein's skepticism about quantum mechanics, there is perhaps some justice in the tension between general relativity and quantum mechanics. In a letter to Max Born, commenting on Born's probability idea, Einstein famously wrote: "Quantum mechanics is certainly imposing. But an inner voice tells me that it is not yet the real thing. The theory says a lot, but does not really bring us any closer to the secret of the 'old one.' I, at any rate, am convinced that He does not throw dice." (Einstein, who said this in 1926, used a gender-specific reference to God typical of the time, but he didn't actually subscribe to a conventional religious view of the deity.)

In the early days of quantum mechanics, Einstein tried to turn his unease into a sharp critique that would demolish the theory. He repeatedly challenged Niels Bohr with hypothetical "thought experiments" that seemed to show that quantum theory and Bohr's interpretation of it did not make sense. The questions Einstein asked were often tough, but Bohr, sometimes after a prolonged period of thought, invariably found a way to resolve each paradox. Years later, John Bell translated one of these thought experiments (essentially the Einstein, Podolsky, and Rosen, or EPR, paradox) into a proposal for a real experiment. This experiment has been performed repeatedly and has confirmed the quantum rules. Einstein's unease is not enough to dethrone the quantum-based view of reality.

But general relativity creates new areas for conflict. Resolving this clash over whether to include probability calculations in the laws of nature may point to answers to some of the toughest scientific questions there are.

As we have engaged with questions about black holes, cosmology, and the cosmological constant, we have been concerned with gravity as a classical theory. This is good enough for the experiences of our day-to-day lives, and good enough for the cosmos. In fact, it is hard to conceive of an experiment that would involve general relativity and quantum mechanics in an essential way. Most of humanity's encounters with quantum mechanics have been in the world of atoms and things smaller, and we have seen that gravity is totally unimportant for such systems. To put this more starkly, just as the quanta of the electric and magnetic fields are photons—discrete bits of electricity and magnetism—so the quanta of the gravitational field are particles called gravitons. For more than a century, scientists and engineers have been able to work with individual photons. At the moment, we can't conceive of, much less perform an experiment to study, a single graviton in isolation. So we are definitely headed deeper into the realm of thought experiments. But let's take a moment to make clear just how remote the quantum study of gravity is from feasible experiment.

Real Experiments for Classical Gravity and Thought Experiments for Quantum Gravity

In making sense of quantum mechanics, one of Bohr's guiding principles was that, in appropriate circumstances, quantum mechanics should look classical. For light, it is the statement that when a ray contains lots of photons, it should obey the rules of classical physics; the discreteness of quantum mechanics should disappear. Light as we experience it day-to-day is a collection of large numbers of photons. A 60-watt bulb gives off about 10^{20} photons every second (a more efficient light gives off more like 10^{18} photons). With this many photons, the light obeys to a very good approximation the rules of classical physics—Maxwell's equations, with no quantum modification.

But with present technology, it is not hard to create situations where one can study one photon at a time. These single photons obey all the rules of quantum mechanics. We can watch a single photon knock an electron off an atom, and test quantum predictions for the probabilities of various outcomes. We can observe one electron collide with a proton, producing a single photon, or two, or three.

But, again, when we study situations with large numbers of photons, electrons, and protons, the quantum effects become unimportant, and Newton and Maxwell's rules take over. Light, for example, is produced when large numbers of charged particles speed up or slow down. We experience light when the electric and magnetic fields in those waves accelerate charged particles in our retina. Einstein understood that, in a manner similar to Maxwell's theory, his general relativity would lead to production of waves of the gravitational field—gravity waves— when mass or other forms of energy speed up or slow down. These waves, in turn, would accelerate matter as they passed by. Because the gravitational force is so much weaker than electricity and magnetism, these effects would be minuscule, even when huge amounts of mass are involved. While there were some prototype experiments in the second half of the twentieth century designed to detect these waves, they had no chance. The first experimental program with any real hope to detect gravitational waves began in the 1990s, and was known as LIGO, for Laser Interferometer Gravitational-Wave Observatory.

Electromagnetic waves, as they pass through a material, jiggle the charged particles in their way, and this allows us literally to see with our eyes, and also gives rise to all sorts of other phenomena. Gravitational waves, on the other hand, jiggle every-thing they pass but hardly at all, so they are extremely tough to detect. Those who conceived the LIGO program asked what might be the biggest source of gravitational waves. The more mass and the faster the objects involved, the stronger the wave. Two colliding neutron stars or, better, two colliding black holes offered what seemed the only hope. These are typically heavy objects, with the mass equal to that of the sun or more; in the seconds before they collide, they rapidly speed up, reaching speeds close to the speed of light. These would be the ideal can-didates, producing the strongest waves one might hope to see. In Einstein's theory, the change in the gravitational field is a distor-tion of space-time. As waves pass by an object, they stretch the space around it. The object appears slightly longer and then slightly shorter, and then slightly longer again. It is this stretch-ing and shrinking that would be the key to detecting such waves.

Now, when I say slightly, I mean *slightly*. The LIGO gravitational-wave detectors are long metal tubes each 4 kilome-ters long. The passing waves from colliding black holes stretch and shrink the space occupied by these huge bars by about 10^{-18} cm, an amount 10^5 times—100,000 times—smaller than an atomic nucleus. Put another way, as a fraction of its length, each bar changes by about a trillionth of a trillionth of its length. Measuring such a ludicrously small change sounds like magic. When LIGO announced its first discoveries, I was teaching a course on general relativity for undergraduates and wanted to explain how the experiments work. I wandered the halls of my department, asking questions about how this was done. No one could give me a complete answer. I had to pore through papers and articles online (and, at the risk of my professional pride, I acknowledge watching an assortment of YouTube videos) to sort it out. What's crucial to making the experiment work is a set of properties of ordinary light, and the use of high-powered lasers.*

I just mentioned the number of photons that emerge from an ordinary light bulb. There is, from these colliding black holes, a far, far larger number of gravitons passing by the LIGO detectors. From this whole collection, one can barely find an observable signal. Divide this signal into something like 10^{70} parts to try to detect a graviton—truly hopeless. Quantum gravity, for the foreseeable future, lies purely in the realm of unrealizable thought experiments.

The LIGO experiment was first conceived in the 1990s and funded by an act of the US Congress. It actually is two experiments, located at two sites, one near Hanford, Washington, one near Livingston, Louisiana. The two sites are intended to provide confirmation for any observed signal. Along the way, there were many challenges. The instruments are extremely sensitive. Even trucks passing at some distance from the experiments produced vibrations that could fake the effects of gravitational waves. In the early phase, an earthquake near the Hanford site required much reworking of the detectors. Ultimately, as a technological accomplishment, the results were amazing. Still, in its prototype phase, the experiment, already

---

*For those who want a little more detail, the experiment uses the interference of laser beams along the two arms. The slight stretching of the bars leads to a slight difference in the length traveled by the light beams moving along the two arms. They arrive not quite in phase with each other. The high intensity of the laser is crucial to obtaining a detectable signal.

large and extremely sophisticated, wasn't really sensitive to the signals of interest. But over the last decade, gravitational waves have been observed from collisions of neutron stars and black holes. A whole new way to study the universe has emerged. Kip Thorne, a theorist, Barry Barish, by background a high-energy experimentalist, and Rainer Weiss, an atomic physics experimentalist, drove this Nobel-winning effort, with a combination of vision, technological innovation, and management skill.

Thought Experiments and Black Holes

Setting aside the practical obstacles to doing any experiment to detect quantum effects of gravitation, we can still ask whether such tests are possible, even in principle. In trying to apply the rules of quantum mechanics to general relativity, physicists have encountered two problems.

In saying that gravitons hardly interact at all, there is a catch. How much they interact depends on their energy. Roughly speaking, if we make the energy 10 times larger, the probability of interacting goes up by a factor of 100. If each particle in a beam had Planck scale energies, the chances of interaction would be reasonably high. An accelerator with as many particles as the LHC, each with this much energy, would require a power source equivalent to about 10^{16} nuclear reactors—almost the power output of the sun—and would still not produce a graviton collision in trillions of years of operation. Based on this sort of reasoning, Freeman Dyson suggested that uncovering a graviton might be impossible *in principle.*

But what powers our thought experiments is the uncertainty principle. While we may not be able to produce very high frequency (energy) in the lab, the uncertainty principle says we can't say there is no such gravitational field for extremely short periods of time. In fact, as a result of the presence of these very strongly coupled, high-energy gravitational fields, the quantum theory of general relativity is out of control. The sort of tricks to calculate with photons, gluons, and other particles of the Standard Model, devised by the likes of Feynman, Schwinger, Tomonaga, and 't Hooft don't work. When one tries to compute the effects of quantum mechanics on general relativity, one ends up writing down expressions that make no sense. Among the first to note this fact, and to try, unsuccessfully, to tame it was Richard Feynman, and many others have attacked the problem through the years. With general relativity and the Standard Model alone, nothing has worked. Some held out hope that this problem was just technical in nature, perhaps like the problems of the Standard Model before the work of 't Hooft and others. However, Hawking pointed out a problem for gravity and quantum that, at least at first sight, appears to be very different and would seem to doom any marriage of quantum mechanics and general relativity. The problem is connected with black holes.

When Einstein wrote down general relativity, he struggled to find experimental tests for the theory. The problem is that most of the time, the theory makes almost exactly the same predictions as those of Newton's theory; the corrections coming from the general theory are extremely small. To find any observable effects at all, Einstein needed to take advantage of the large mass of the sun and study phenomena close to the sun. Even then, the effects were very tiny, just barely large enough to be measured with the technology available in the early twentieth century. But there are situations where the effects of general relativity are important, even overwhelming.

Shortly after Einstein wrote down his theory, Karl Schwarzschild, a German physicist and astronomer serving in the army during the First World War, found a solution of Einstein's equations which describes the gravitational field outside a star (sadly, Schwarzschild died not long afterward). The *Schwarzschild solution* readily reproduces Einstein's result for the bending of light by the sun and for the precession of the perihelion of Mercury.

For a sufficiently heavy, small star, this solution describes a black hole. Consider a space traveler headed to the black hole. At a certain distance from the center of the black hole, known as the *Schwarzschild radius*, strange things happen—the stuff of a horror movie. Time and space seem to change roles. Light rays coming from nearer the star turn around and go back— the gravity is so strong that they can't escape. For the sun, the Schwarzschild radius is about a kilometer, so it lies well within the sun, where the solution isn't applicable. But we know that there are much smaller, denser objects that might lie inside their Schwarzschild radii. We've said that neutron stars, for example, have masses comparable to that of the sun and radii of about a kilometer. So we can readily imagine slightly smaller or slightly more massive objects that *are* black holes. For such objects, the Schwarzschild radius would be like the horizon we see when we look out at the ocean. Watching from a distance, objects approaching the black hole would disappear from view as they passed through this black hole horizon.

While we've seen that black holes are real, they also turn out to provide an extremely interesting set of thought experiments; we might think of them as a theoretical laboratory. One of the striking features of black holes in classical general relativity is that they are almost featureless. If you know their mass, their electric charge, and how fast they spin, you know everything you can possibly know about them. They may have arisen from the collapse of a complicated star, surrounded by planets with advanced civilizations, but when they formed, all of that information simply vanished. This is different from a fire or an explosion. There, you might hope, with a huge amount of work, to reconstruct all the original information by looking through the ashes and the outgoing radiation (light and heat) from the explosion. For the collapse to a black hole, this looks to be impossible, at least classically. For a classical physicist, this may seem puzzling, but it does not, by itself, destroy his or her theoretical framework.

But, early on, there were hints that in a quantum theory, black holes could not behave this way. The first physicist to raise questions about the classical picture was Jacob Bekenstein, a theorist at the Hebrew University in Jerusalem. He noted an analogy between black holes and the second law of thermodynamics. The second law says that entropy—which is a measure of disorder we encountered in our discussion of baryogenesis and inflation—always increases. For black holes, there is a quantity that always increases: the area of the black hole horizon. Whatever one does to a black hole—throwing in tables, chairs, planets, other stars—the mass increases and the area of the horizon increases. Bekenstein proposed a precise relationship between the black hole area and the entropy and suggested that black holes were actually thermodynamic systems with a temperature.

But what could this mean? Generally, we think of temperature as measuring the typical energies of some set of particles—atoms, molecules, photons. But we've said that from the outside, we have no information about the black hole apart from some gross properties such as its mass, and we certainly can't identify things like particles. It was Hawking who discovered the sense in which black holes have a temperature.

Hawking, for almost fifty years, was a leading figure in the study of general relativity. He was also remarkable in bringing the subject to the attention of the general public. Born and educated in England, he developed amyotrophic lateral sclerosis (ALS), also known as Lou Gehrig's disease, when he was still a student, in 1963 at age twenty-one. This is a generally fatal neurodegenerative disease. When first diagnosed, he was not expected to live more than a couple of years, but he survived and remained active, while significantly disabled, until 2018. He was determined he would not live a life defined by his disabilities, and this is reflected in his scientific accomplishments and in his complex personal life. He clearly enjoyed his celebrity and would exploit it in ways that colleagues and others found alternatively charming and irksome. He did not hesitate to pontificate on science, politics, and religion, and he was notorious for scientific bets and pranks.

Early in his career, with Roger Penrose, he studied aspects

of Einstein's equations important to understanding the universe. Einstein's equations cease to make sense as one approaches the earliest times, the moment of the big bang. Hawking's work with Penrose established that this is a general feature; it can't be fixed by tweaking the history of the universe a bit. I've mentioned the 2020 Nobel Prize in Physics, awarded to Andrea Ghez and Reinhard Genzel for their discovery of the black hole at the center of the Milky Way; Roger Penrose was the third recipient, for his work proving that black holes inevitably have a singularity at their center in Einstein's theory.

Hawking's most notable scientific contributions relate to black holes. He showed that black holes are not really black; they give off radiation. How does this happen? In quantum mechanics, the uncertainty principle permits brief violations of energy conservation in ordinary space-time. As a result, for an extremely short time, a particle and its antiparticle can appear, even in a complete vacuum, i.e., with no other sources of energy around. These particles will then annihilate and disappear again. In flat space, this is of no observable consequence; we need to probe the vacuum with a source of energy to have physical particles appear. But near the horizon of the black hole, one of these virtual particles can escape, while the other falls back in. The escaping particle can borrow some energy from the large gravitational field, so energy is still conserved. But now there is some radiation, and the total energy—mass— of the black hole is slightly depleted. Hawking found that the emission is just that of a black body of the type we have discussed for the cosmic microwave background. The temperature of the black body is exactly that anticipated by Bekenstein.

So quantum mechanically, the black hole appears a much more complicated object in a quantum world than in a classical one. There's a lot going on inside. The black hole is not static. It gradually evaporates, eventually disappearing altogether. For a solar mass black hole, the time for the entire object to evaporate is very long—about 10^{67} years, far, far longer than the present age of the universe. But we can contemplate smaller black holes, which might be disappearing today. At the end of their lifetimes, there would be a large burst of energy. Astrophysicists search for this possibility. But we'd have to be quite lucky to find such a thing, and so far, there is no evidence for black holes of this size.

His discovery propelled Hawking to scientific fame and to prominence in the popular mind. He eventually came to hold Isaac Newton's position, Lucasian Professor of Mathematics at Cambridge University. There is perhaps some irony in that the position was created to protect Newton, who was a religious Christian but an adherent of heretical views, while Hawking played up his own image as a proud atheist.

Hawking's Thought Experiment

Hawking's (theoretical) discovery of the Hawking radiation was a major accomplishment. But in 1976, he proceeded to consider a thought experiment, which has puzzled theorists ever since. To understand the problem, we have to consider another aspect of quantum mechanics, having to do with information. As discussed, quantum mechanics deals in probabilities. Probability is a useful concept, and it can sometimes be complicated, but there is one aspect so simple as to be obvious and so obvious that its importance is easily overlooked. It helps to think about a case many of us encounter in some form. If you enter your state or national lottery, you focus on your chances of winning. If you buy one ticket and there are 10 million lottery tickets sold, your chances of winning the jackpot are 1 in 10 million. That's a really tiny number. To give my students a sense of such things, I encourage them to think about such numbers by comparing their chances of being killed in a traffic accident on a given day. There are about 100 traffic fatalities in the United States per day. With 300 million people, in round numbers, this means that any person's chances of dying in an accident are about 1 in 3 million. I explain my attitude—this is such an unlikely event that I don't need to worry that it will be my fate. But if the chances of winning the lottery are as small or even smaller—well, that can never happen! (My apologies to state lottery directors who might be reading this.)

But a simple fact about probabilities is that the probability that *something* happens is 1. I either win or lose the lottery; I am killed in a car accident today or I am not. One of the most important features of the Schrodinger equation is that it has this property. It's less obvious than for the lottery. To see it, you have to study the equation with some rather sophisticated mathematics. But if this were not true, the probabilistic interpretation of quantum mechanics would not make sense.

This fact about probabilities is related to the question: Can information disappear? Of course, we all forget things, lose records of various types, or deliberately shred or burn papers that might embarrass us or expose us to risk. But we believe that with enough patience and resources, we could reconstruct this information. In our computer age, we are used to thinking about and measuring the *quantity* of information. The hard drive on my laptop holds 80 gigabytes of information. My internet connection delivers so many bytes in a second. If I had unlimited resources, again, whether my laptop died or my house burned down, I could reconstruct this information. The amount of information in a system (or the universe) doesn't change, though much of it may be hard to access. Before including general relativity, the same is true in quantum mechanics. The information about the state of a quantum system is encoded in the now famous *Schrodinger wave function*. For a complicated system, like a collapsing star, there is a lot of information—an unimaginably large amount. In classical physics, there would be the positions and velocities of all the nuclei and electrons. In quantum mechanics, there are complicated relations between all of them; one can't give the probability that one particle is at a point without specifying also the probability of finding all the other particles at particular places as well.

So a collapsing star contains a *huge* amount of information. Thanks to Hawking, we know that if the star is heavy enough, it forms a black hole and then slowly evaporates, emitting radiation. All of the complicated structure in the initial star has turned into the vanilla radiation of a warm body. Where did all of the information go? Hawking, in his 1976 paper, argued that the information was simply lost. Quantum mechanics, he asserted, breaks down near black holes.

This was a thought experiment par excellence. It threatened

either quantum mechanics or general relativity or both; and it was truly a thought experiment, involving questions so extreme that it was hard to conceive how they could be attacked experimentally. Many leading theorists were drawn into the debate over this question. Some thought, indeed, one has to redo quantum mechanics or general relativity. Others were more skeptical. Perhaps, for example, the evaporation of a black hole is like a lump of ash, the burning of a log in a fireplace. Surely the laws of quantum mechanics don't break down when an object burns. In that case, the resolution of the puzzle is that the outgoing radiation is not exactly that of a black body; there are subtle connections between the outgoing photons. But it turned out that the answer to Hawking's question about the black hole problem could not be so simple; the structure of space and time makes it hard to understand how such correlations might arise. There were other proposals, none very satisfying. Perhaps Hawking was right: Quantum mechanics or general relativity had to give.

But it turned out that there was a situation where black holes should exist and quantum mechanics should make sense: string theory. String theory has provided at least a partial resolution of the puzzle. So let's dig in.

String Theory

Why do some physicists view string theory as a sort of holy grail, even calling themselves string theorists, while others view the subject with total disdain, often wearing their ignorance of even its most rudimentary aspects as a badge of honor? We have already had a brief encounter with string theory and noted that it somehow builds in general relativity. The rather peculiar history of the subject gives some insight into the fascination that strings hold and how general relativity and a world that, at least very roughly, can resemble ours emerges, so let's explore how physicists stumbled on these intriguing theories. In the late 1960s, before the emergence of quantum chromodynamics (QCD), the strong interactions were a source of great frustration for theorists and experimentalists. There were literally hundreds of different types of particles. The quark model attracted great interest, but it was not clear—even to Murray Gell-Mann himself, that quarks were real entities. In the quark model, the different strongly interacting particles were built up of quarks, arranged in different ways, much like, in Niels Bohr's picture of the atom, electrons can have different orbits. We refer to this as a composite picture, where the different particles were excited states of simpler ones. Perhaps, some reasoned, there should be a more democratic picture in which all the hadrons were on a similar footing. Playing with models of this sort, Yoichiro Nambu and Leonard Susskind came up with the idea that the strongly interacting particles could be modeled by strings.

When we say strings, we mean lines as opposed to points, as noted earlier. But why did Nambu and Susskind think this had something to do with the zoo of strongly interacting particles? Working out what such a theory does is not so easy, but it is a problem that would be a suitable homework assignment for a hardworking graduate student. Homework problem 1:

Work out the theory of a classical string, and require it to obey the rules of special relativity. Homework problem 2: Apply the rules of quantum mechanics to the results of problem 1.

Working on the first problem, the student would discover things we all know about the strings of musical instruments. Like the strings of a guitar, violin, or piano, strings can vibrate at a characteristic frequency, the *fundamental*, but also at higher frequencies—any whole number (1,2,3, . . .) times the fundamental. These are the *harmonics*. They are crucial to how we hear music. One important difference: Unlike the strings of a musical instrument, which are tied down, Nambu and Susskind's fundamental strings could also fly around as a whole.

Problem 2, including quantum mechanics, at first sight would be a disaster. Like the problem of putting a square peg in a round hole, the student would find that the rules of special relativity worked but not those of the quantum theory, or the other way around, at least in our usual world with 3 dimensions of space and 1 of time. Sensible students would give up at this point, but one not grounded in reality might contemplate a world with more dimensions of space. This one would find that with 25 dimensions of space (26 of space-time), one could satisfy the demands of Einstein and of Schrodinger-Heisenberg-Dirac.

This oddball student might push on. She would realize that frequencies translate into energies, and that in Einstein's relativity, energies translate into masses. The different ways a string can be excited, in other words, look like particles of definite mass. And there is an infinity of ways that strings can vibrate—an infinite number of harmonics. So there would be an infinite number of possible particles, of higher and higher masses. This was how things seemed to work in the strong actions.

Even allowing, somehow, for this weird space-time, some things were badly off for a model of the strong force. There was a frightful particle called a *tachyon*. Such a particle would move faster than light and would lead to other, nonsensical results. That was bad enough, but there was a second problem, a particle with *zero* mass. There is no candidate for this particle in the strong interactions. Another weird feature of this particle was its spin. It wasn't zero, like the π mesons, or one, like the photon, but spin two.

With further work, things only seemed to get worse. The original string model didn't have any particles with spin-1/2, like the electron and the quarks. When the theory was modified to include these, one sometimes got rid of the tachyon, and the theory now made sense in *9* space and 1 time dimension, but at a still bigger price: There was a strange symmetry between particles of different spin. The student had discovered supersymmetry, but if she simply wanted to model the properties of the hadrons, this was just one more unwanted development.

Finally, I should say that our poor graduate student was finding the homework, by this stage, next to impossible. The calculations were very hard and intricate. And by now the payoff seemed extremely limited. There were much easier calculations to do in QCD, the new theory of strong interactions, and the reward was immediate—and successful—applications to experiment.

People trying to match these theories to the observed properties of the strong interactions were put out of their misery by the discovery of QCD in 1973. Starting my own graduate studies in 1974, I felt I had been spared the wrong turn of string theory and was bewildered that there were a few theorists who wouldn't let go.

But I now see those who persisted as visionaries. In fact, there were a small number of physicists who believed that the structure of string theories was so remarkable that they must play some role in nature. Joel Scherk and John Schwarz, in 1974, made a bold leap. They noted that the graviton of the would-be quantum version of Einstein's theory has spin two. They suggested that, in fact, string theory was a theory not of the strong interactions but of general relativity.

This was particularly weird. Our poor graduate student was not told to implement the principles of general relativity, only special relativity. She was not due to take the GR course until the following year. Yet she seems to have achieved what Einstein required, albeit in the wrong number of dimensions. In fact, earlier work, especially by Steven Weinberg, had shown that any sensible quantum theory of a massless spin-two particle necessarily incorporated Einstein's principle. So provided string theory was a sensible theory, it had to work this way.

Now, what got Scherk and Schwarz so excited is something our graduate student perhaps did not appreciate. We have said that the quantum theory of general relativity is subject to horrible problems at very high energies. The calculations one tries to do, using the rules that Feynman laid down, are nonsensical. But string theory is different simply because lines (as opposed

273

to points) are not prone to certain mathematical problems that make the calculations nonsensical. Scherk and Schwarz argued that, as a result, string theories might deliver sensible quantum theories of gravity.

Still, all of this looked a bit crazy. What was one to make of the 26 dimensions (25 dimensions like our 3 of space, and 1 of time), and the tachyon of one version of the theory? Or of the 10-dimensional version, without a tachyon at least, but with other weird stuff. At best these looked like nice mathematical models, seemingly having nothing whatsoever to do with the real world.

But Scherk and Schwarz were undaunted. They suggested that maybe the extra dimensions of space were not on the same footing as the usual three. Maybe they're there, and we just don't see them. More precisely, the two suggested that in these other dimensions, space might be curled up, into something like very tiny circles.

The idea that space has more than three dimensions, the others being tiny, has a long history. It was proposed early in the twentieth century by Theodor Kaluza and Oskar Klein and was of great interest to Einstein. In fact, it was one of the ideas Einstein explored in his search for a unified theory of general relativity and electromagnetism.

What is this idea of extra dimensions, and what were Kaluza and Klein after? The first thing to realize is that this is not an idea our brains are wired to contemplate. Don't even try to make a picture in your mind of what a world of 4 or 5 spatial dimensions looks like, never mind 25. Instead, just accept the notion of these extra dimensions as an abstract bit of mathematics. It's an extension of ideas many of us encounter already in high school mathematics. The seventeenth-century French mathematician Rene Descartes taught us to think of the plane in terms of a chart, or map, with coordinates labeled by x and y. To give the location of a point in the plane, one gives the value of x and y, say x=3, y=4, written in a compact way (3,4). Descartes's idea generalizes to 3 dimensions, with coordinates x,y,z, or a triplet of numbers, e.g., (1,3,5) to designate the location of a point in space. Now, while we can't picture it, mathematically we can keep going. That is, we can have 4 dimensions, (x,y,z,a) or five (x,y,z,a,b) and so on. These extra dimensions, the fourth or fifth, might be infinite, but they also might be finite, like circles or spheres.

Kaluza and Klein worked precisely in this abstract framework, with one extra dimension. They took this dimension to be a small circle, though at the time they didn't really have a way to think about how small these dimensions should or must be. But they made a discovery that they found very exciting. In their basic, higher dimensional theory, they included general relativity. But from the perspective of an observer in four dimensions, what emerged was general relativity *and* electromagnetism. This is why Einstein was so excited. Somehow, the higher dimensional general relativity bequeathed a unified theory of general relativity and electricity and magnetism to four dimensions. In any case, Scherk and Schwarz, confronting the extra-dimensions problem of string theory, suggested that this sort of *compactification* of extra dimensions might provide the solution. All those extra dimensions, perhaps, were small, compact spaces, so small that their effects would be essentially unobservable; space-time would appear four-dimensional, perhaps with some new particles and interactions inherited from the higher dimensions. If you accepted this, the other good features of string theory survived. The theory was turning into a framework radically different from anything proposed before. String theory might be a theory of gravitation and other interactions, without the diseases of four-dimensional general relativity because of these differences.

Not everyone jumped on the bandwagon—there was no bandwagon. At that moment in time, QCD was new and exciting. Quantum general relativity was not on most physicists' front burners. Joel Scherk died in 1980 when only thirty-four, but John Schwarz persisted in a collaboration with Michael Green (then of Queen Mary College, London, followed by several years at Cambridge University, where he held the Lucasian professorship previously held by Newton and Hawking, among others; and most recently at Queen Mary University of London). They began a long and arduous program of study of the supersymmetric strings or superstrings, seeking to establish that the theories really were sensible theories of gravity, and developing techniques for doing calculations. They made great progress. But an obstacle came from a different quarter.

Edward Witten was among the few following these developments with great interest. As a postdoc at the Institute for Advanced Study, I remember lunch conversations about string theory. Edward told us that Green and Schwarz were up to something really important, but also very hard. He could see serious obstacles to the program. While encouraging us to work on the topic, he warned that, although interesting and perhaps ultimately providing a complete underpinning for the laws of nature, this problem might be similar to problems in mathematics that required many hundreds of years for their resolution. My younger colleagues and I smiled politely and went back to our offices to continue working on whatever occupied our attention at the moment. Thousand-year problems might be interesting, but they are not usually good career moves. And I have to say that, in my encounters with John Schwarz in those days, I found his persistence bewildering. The problem was so hard, with no promise of yielding anything of interest in the foreseeable future. Indeed, at the time, John had only some sort of tenuous position at Caltech. I wondered why he would devote all his energies to a problem so far from the mainstream.

Ed's view of the challenges of this theory arose in part from its mathematical complexity. But he had some concrete, conceptual concerns that appeared to raise a possibly insurmountable obstacle for string theory. These were of two basic types. One had to do with finding in string theory the particles and fields of the Standard Model. Another had to do with potential diseases of the theory—in other words, anomalies.

Witten had been fascinated for some time with the Kaluza-Klein program. He was troubled, though, by the lack of clear rules for how this might work. There seemed to be an infinity of possibilities. The number of extra dimensions might be one, two, three, six if the underlying theory was string theory, but there were also hints of a possible role for seven. And what might these extra dimensions look like? Six circles? A sphere if there were two extra dimensions? Much more exotic objects in more dimensions? Witten realized that, independent of these hard-to-pin-down details, there were some minimal requirements for this compact space. Getting enough particles to account for the gluons, photon, W's, and Z didn't look too hard. One could get particles like electrons, quarks, and the like. But some of these requirements looked to pose a serious stumbling block to obtaining the Standard Model in the Kaluza-Klein program. The Standard Model has a striking property, which we encountered before: It violates parity, or, as we have seen, it distinguishes an event from its mirror image. What Witten realized is that if the world we see originates from the compactification of a higher dimension, it is hard to understand this fact. Using some very sophisticated mathematics, he was actually able to prove that one could not obtain the Standard Model with its striking feature of parity violation from only general relativity in a higher dimensional space-time.

So the Kaluza-Klein program, which so beautifully seemed to account for the sorts of interactions that appear in the Standard Model, appeared to have hit a brick wall. With Witten's result about parity, the simplest possible way out was to give up this remarkable feature and accept that the Yang-Mills theory of the Standard Model was already present in the higher dimensional theory. In terms of particles, this means the theory should contain photons, Wbosons, and gluons—the *gauge bosons* of the theory. Of the superstring theories classified by Green and Schwarz, one class had extra gauge fields, like those of the Standard Model. So if one was willing to accept compactification of extra dimensions, without demanding that the Standard Model emerge from the higher dimensional general relativity, one could perhaps find an acceptable realization of the physics we know.

But the second objection Witten raised to the higher dimensional program, concerning an obstruction called *anomalies*, seemed insurmountable. Already in four dimensions, not every classical field theory one can write down makes sense as a quantum theory. The problems are similar to those we talked about in string theory in dimensions other than 10 or 26: The probability interpretation of quantum mechanics can't be sustained. Even the Standard Model just barely avoids these anomalies, slipping past the pitfall in a rather intricate way. But in higher dimensions, these conditions are much more restrictive, and with general relativity in the picture, Witten showed that these were almost impossible to satisfy. In fact, he proved that all but two of the known superstring theories suffered from anomalies. This appeared to mean that only two of the theories made mathematical sense. The two had no particles that could play the role of the gauge bosons of the Standard Model. It appeared, to many of us, to be the death nell of string theory. I have to confess that, at the time, I was awed by Witten's arguments and relieved that, once again, I had been spared any obligation to work on string theories.

But once more, Green and Schwarz refused to give up. While Witten's arguments appeared quite general, they were convinced that their string theories would somehow evade them, and they set out to compute the anomalies directly in superstring theory. They found something remarkable. Witten was *almost* right. Of the infinite collection of theories, three were mathematically consistent. Of the three—just one—exhibited gauge symmetries. These symmetries were big enough to encompass those of the Standard Model. Now they had a puzzle, though. Why did Witten's arguments almost always work? It should be said that the history of physics is full of theorems that prove such and such can't be true. Such theorems are called "no-go theorems." Often these theorems turn out to have loopholes. They are not really wrong, but there is sometimes an assumption, not clearly stated or whose significance is not appreciated, that permits exceptions. Something like that had to be going on with the anomalies of theories in 10 dimensions, and Green and Schwarz set out to find it. They dug deeply into their calculation and Witten's and found the loophole. But they also found another surprise. There was one other possibility for a theory with gauge symmetries and no anomalies. At that time, no string theory with this structure was known.

The effect of these discoveries, made in the summer of 1984, was electrifying. They launched what has come to be called the *First Superstring Revolution.* Perhaps because the difficulties posed by Witten had seemed insurmountable, it now appeared to many that we might be on the brink of discovering some sort of ultimate theory. I remained resistant but started to worry that I was in danger of missing out on something big. My colleague at the Institute for Advanced Study, Nathan Seiberg, pressed me to take the plunge. Nati was quite young at the time, relatively recently arrived at the Institute from Israel. We had spent the previous year in a successful collaboration on problems in supersymmetry with Ian Affleck. We'd also each had a child during that period. Now Nati dragged me, more or less kicking and screaming, into string theory. The subject was difficult and the existing literature obscure. But we were helped that fall when David Gross—the same David Gross who was so critical in the development of QCD—taught a course on string theory at Princeton University. David had worked on string theory in his graduate student days and was quickly coming up to speed on the recent developments. The class was full of graduate students but also a slew of famous Princeton faculty, all frantically taking notes and doing homework. But meanwhile, David and collaborators Jeffrey Harvey, Emil Martinec, and Ryan Rohm were building a string theory of a new type, which filled out the remaining possibility discovered by Green and Schwarz. The details of this theory, which they called "heterotic," are rather technical. It involves a branch of mathematics familiar to many physicists, known as group theory, but a particular corner of this mathematics that was foreign to most. These theories had features promising the embedding of the Standard Model in a structure including general relativity as well.

There remained the problem of extra dimensions. All the superstring theories were theories of 10 dimensions. Then, in early 1985, Witten in collaboration with Phil Candelas (now at the University of Texas), Gary Horowitz (UC Santa Barbara), and Andy Strominger (Harvard) found that the equations of the string theories were solved if the theory had 4 large dimensions, like those of our ordinary experience, and 6 small ones. The 6 small ones were a space of a particular type, known as a Calabi-Yau space, named for the two mathematicians who had first discovered and explored them. Wonderful results followed. The gauge theory of the Standard Model emerged quite naturally. Supersymmetry, in the form that had been advocated as a solution of the hierarchy problem, also emerged, almost automatically, it seemed.

This discovery also addressed another long-standing puzzle of the Standard Model. We have seen that there are three generations of quarks and leptons. This repetitive structure is puzzling. It would seem we could have a perfectly fine universe with just the first generation. Indeed, when the muon was discovered (or, more precisely, identified) shortly after the Second World War, the theorist Wolfgang Pauli asked "who ordered that." Various ideas had been put forward through the years, but none were persuasive. The compactified string theories produced this bizarre repetitive structure almost all the time. If anything, there was an embarrassment of riches. More often than not, one found *too many* generations.

I was still resistant. But a conversation with Witten one Friday afternoon altered my thinking. Being skeptical and even somewhat annoyed by these developments, I asked Ed how, in this framework, he would solve one of the long-standing problems of grand unified theories, closely tied to the hierarchy problem. He frowned and said he didn't have a good answer. I was feeling smug, my lack of interest justified.

I ran into him the following Monday, before David Gross's class, and he took me aside and said, "By the way, I have an answer to your question . . ." He took me to a blackboard in a hallway in Jadwin Hall, the physics building at Princeton, and explained. I didn't completely understand, but I understood well enough to get the main point. Now I felt I'd better get on this bus—all the basic questions that troubled me in physics, it seemed, were about to be answered.

A Roadblock to Relating String Theory to Nature

There followed several months of very intense work and progress, to which I made some contributions. But Nathan Seiberg and I also realized that there was a fundamental obstacle to relating string theory to nature. It has to do with whether one can actually figure out what the theory predicts, if anything. The problem has come to be known as the Dine-Seiberg problem. To appreciate the issue, it is helpful to return to quantum electrodynamics (QED), the theory that was so well understood starting with the work of Feynman, Schwinger, and Tomonaga. There, for whatever you want to calculate, is a simple first stab you can take, and it's usually pretty easy, in the sense that you can give the problem to a good graduate student and they should be able to come back with the answer in a few days. Then there is a more accurate calculation you can do. Typically, the correction to your original calculation is only a small correction, about a part in a thousand. This may take the graduate student a few months. You can do a still more accurate calculation—now good to a part in a million. This may take an experienced professor and a collaborator a few years. And so on. The reason each correction is small is that there is a small quantity in QED, the fine structure constant, whose value is famously about 1/137 and is customarily denoted by the Greek letter α. The first, simple calculation involves one power of α. The next, harder computation involves two powers of α, and is about 1,000 times smaller than the first, and so on. In the weak interactions, there is a corresponding quantity whose value is about 1/30. For the strong interactions, the quality of these approximations depends on the energy, but for Higgs production at the LHC, for example, the small quantity is denoted as α_s and is about 1/10. You don't need to remember anything about these numbers except that they are small, and it is their smallness that allows accurate predictions.

String theory also has such a quantity, usually called the string coupling constant, and denoted by g_s. For small values of g_s, one can calculate, and the smaller g_s the more accurate the calculations. The calculations are harder than in QED or the Standard Model; indeed, much of the early progress by Green and Schwarz was in developing rules for these computations similar to Feynman's.

One of the really remarkable and attractive features of the theory is that the theory itself should determine the reality in which we find *ourselves*. In the Standard Model, there are three numbers like the fine structure constant that we have to go measure. We don't have an a priori theory for what they should be. In grand unified theories, this is slightly better; one coupling is determined, but you still have to go out and measure the other two. But string theory is different. The couplings are determined by the theory itself. In fact, every quantity— the masses of the quarks and leptons and the Higgs particle, the value of the cosmological constant, everything else we might find in a table in the back of a textbook (or searching on the internet) should emerge from the theory itself. This is one of the reasons string theory is viewed as a possible ultimate theory. The problem is that there is no reason why there should be a small quantity that would allow us to easily calculate anything. One might expect that the analog of the fine structure constant, g_s, should be some sort of universal number, like one or three or π. Seiberg and I made this statement sharp, showing that the dynamics one might expect to determine g_s would fix a value that is not small.

There were other problems, which turn out to be related. One, stressed from the beginning by Witten, is that the dark energy would be expected to be just as large as the naive guesses we spoke of before. No simple answer to this problem has been uncovered. A sensible response to these and other challenges might have been to drop the theory, and this was certainly the viewpoint of many theorists. It was articulated most strongly, and with some humor, by the Nobel Prize winner Sheldon Glashow and Paul Ginsparg, then a faculty member at Harvard and subsequently a MacArthur Prize winner at Cornell for his work bringing scientific publication into the internet age. In an essay entitled "Desperately Seeking Superstrings," they criticized the string program as almost unscientific, given the challenge to actually extract predictions.

But despite these criticisms, string theory is so remarkable, eluding so many obstacles to a unified theory, that work on the theory continued. Further discoveries gave support to the view that physicists have uncovered a structure that is the basis of some ultimate unified theory. Dramatic leaps came in the mid-1990s, under the heading "duality." This *Second Superstring*

Revolution, as it came to be known (the first was the period of rapid progress following the discovery of anomaly cancellations), was launched by a talk by Ed Witten at the annual Strings conference in 1995, held that year at the University of Southern California in Los Angeles. The conference is not a meeting I attend religiously, but I happened to be at the conference that year, which was very fortunate. This was still before the prevalence of PowerPoint, Keynote, and their relatives. The technology of choice for presentations, when not a blackboard, was the overhead projector, with writing by marker on plastic slides. Witten spoke for about an hour, showing over eighty of these transparencies. I, like most of my colleagues, will advise students that if they speak for an hour, they shouldn't show more than about twenty slides, perhaps thirty. Those who show, say, sixty slides usually put their audience to sleep, with little interesting content. Not this talk. Every slide introduced something completely new, and a research program engaging much of the community was launched.

To appreciate the significance of this work, one needs to understand a little bit about the state of the subject immediately prior to the conference. There seemed to be five different string theories, the two Type II theories that had passed Witten's original test, the theory found to be consistent by Green and Schwarz, called the Type I theory, and two heterotic theories discovered by Gross and his collaborators. The differences between these theories were rather striking. Some had gauge symmetries, some not; some had closed strings only—strings that close on themselves—and some had both closed strings and open strings with ends. But the upshot of Witten's talk was that each of these seemingly very different theories was a realization of one underlying theory. One theory compactified on small circles, for example, is equivalent to another theory compactified on larger circles; one theory with a small coupling constant is equivalent to another with a large coupling. When compactified, the two Type II theories are equivalent to one another, as are the two heterotic theories. The type II theories, in suitable situations, are equivalent to the heterotic strings. This web of dualities was supplemented by one more: for extremely large coupling, the Type IIA theory, seemingly in ten dimensions, becomes a theory in eleven dimensions. Eleven dimensions had been known for some time to be special, in that it is the highest dimension in which one can write a field theory with supersymmetry. The theory possesses two-dimensional objects, analogous to strings, called membranes. These membranes, in fact, are strings in waiting. If one dimension is a small circle, and if the membrane is wrapped around the circle, from the perspective of the remaining dimensions, it looks like a string. Witten dubbed this theory in 11 dimensions "M-Theory," where the *M* has been argued variously to stand for Mysterious, Membrane, or Magic. But whatever it stands for, these connections were quite remarkable. They suggested to many that there is only one possible theory of quantum gravity. The various theories, which look so different, are all part of one structure. Perhaps, in some bizarre way, we have stumbled on it. It would seem humans are, at this stage, a bit like the blind men in the parable of the elephant.

Witten's talk in Los Angeles opened a floodgate of activity. Several remarkable discoveries followed.

Let us step back for some perspective on this development. I described before the problem I would give a graduate student: Consider a theory of a string, subject to the rules of special relativity and quantum mechanics. Then I said that all sorts of marvelous things happen. One finds that such a string has massless excitations of spin two, which interact with each other and with matter just as the gravitons of (quantum) general relativity should. But the student—and the professor—have (almost) no clue why this is true. We didn't enunciate some principle from which this follows. Essentially, this dumb idea, of considering objects just the next level up in complexity from points—yields an extraordinary set of riches. But it is troubling that we don't really have a complete description. We just have a set of rules for calculations, and these rules work only when the coupling constant of the theory is extremely small. This is part of what had bothered Seiberg and me: If string theory has something to do with nature, it is only in a regime in which we would seem to have no idea what the theory is. While hardly providing a complete picture, Witten's talk at the conference hinted at aspects of a larger structure.

Witten's web of string dualities, while compelling, had gaps filled with plausible but unsubstantiated conjectures. Joe Polchinski, of the University of Texas at Austin, had previously discovered objects he called D-branes in string theory, which played a role somewhat similar to the magnetic monopoles we have encountered in grand unified theories. Now he recognized that these objects filled in exactly these gaps. A much more compelling picture of the whole structure of string theory had emerged.

Taking advantage of Polchinski's discovery, my good friends Tom Banks (Rutgers and then UC Santa Cruz), Willy Fischler, Steve Shenker (Rutgers and then Stanford), and Leonard Susskind put forward a theory that completely describes 11-dimensional M-theory. It was a system of equations that may be hard to solve but which, put on a computer, *can* be solved with a specified set of rules in a definite amount of time. Even more dramatic was an observation of Juan Maldacena (then at Harvard, now at the Institute for Advanced Study) that under certain circumstances, string theory is *identical* to certain well-understood quantum field theories. The circumstances are string theories in space-times corresponding to universes with a negative cosmological constant (and various numbers of dimensions). These spaces are known to general relativists as anti-de Sitter space-times (after Willem de Sitter of Leiden University), or AdS space-times, and Maldacena's observation is called the AdS-CFT correspondence, for the class of field theories involved. The Second Superstring Revolution continues to this day.

While not exactly solving the problem that Seiberg and I had put forward, these results have provided many insights and have proven useful in other areas of physics.

D-Branes and Hawking's Paradox

In addition to filling in the gaps in Witten's picture of the string dualities and providing the underpinning for the Matrix Model and Maldacena's AdS/CFT correspondence, the D-branes in the hands of two theorists at Harvard, Cumrun Vafa and Andy Strominger, allowed a solution of Hawking's puzzle. The point is that, while their discovery hinged on Polchinski's remarkable insight, the D-branes themselves are actually rather simple. Using them, all sorts of calculations can be done with paper and pencil rather than with powerful computers. Strominger and Vafa realized that certain collections of D-branes are black holes, and that it is not hard to compute their entropy using the rules of quantum mechanics. One finds, exactly, the Bekenstein-Hawking result. This in a theory that obeys all the rules of quantum mechanics and relativity. So there is no paradox.*

Actually, while this result settled the question in an abstract way, it left many physicists dissatisfied. Because the calculation is done in a situation that doesn't much resemble an astrophysical black hole, it is hard to figure out just what went wrong with Hawking's argument. There remains something important about the way general relativity works which we don't yet fully understand. Most recently, this was formulated as the *firewall paradox* by Polchinski and collaborators. It is usually said of the horizon of a big black hole that nothing interesting happens there. If you are in a rocket falling freely in the black hole's gravity, you won't notice the horizon, even though your friend, watching from a distance, sees you disappear and

---

*There are some subtleties in the analysis. The collection of D-branes is not actually a black hole in the regime where the computations are easy. But the result turns out to be true both in the easy regime and in the hard one. The actual computation involves a counting of the number of quantum states of the system of a given energy, which is related to the entropy.

knows you will ultimately meet your fate at the singularity at the center of the black hole. But quantum mechanically, with certain assumptions about the way the Hawking paradox is removed, this can't be true. Polchinski and his colleagues argued that the rocket ship will be explosively zapped by high-energy radiation. But there are other possible resolutions of what has become known as the AMPS paradox, and this remains the subject of heated discussions among those like Banks, Shenker, and Susskind—who are determined to resolve these questions. Tragically, Joseph Polchinski passed away in 2018 as a result of a brain tumor.

In the early days of string theory, there was a widespread belief that the theory was the *unique* theory of gravity. The Second Superstring Revolution reinforced this view; the few known but very different looking theories appeared part of the same theory. But there remains the question: Is there only one possible theory that incorporates general relativity and quantum mechanics? Is what we call string theory just one corner of the space of possible realizations of this theory? This question takes on greater urgency in the framework of the landscape, the subject of the next chapter.

CHAPTER 14

THE LANDSCAPE OF REALITY

String theory is quite impressive. With a very simple set of inputs, for reasons only dimly understood, it yields a structure containing Einstein's general theory and the Standard Model, consistent with principles of special relativity and the rules of quantum mechanics. It doesn't suffer from the infinities anticipated by considering general relativity in quantum field theory, and it resolves, in ways at least partially understood, the issues raised by Hawking.

String theory has other remarkable features. While formulated most simply in 10 space-time dimensions, it is readily realized in 4. Not only can it exhibit the gauge bosons of the Standard Model but also the repetitive structure of quark and lepton generations. All the constants of nature should be calculable within the theory. There's yet more. String theory has structures like those of grand unification, magnetic monopoles, supersymmetry as anticipated to solve the hierarchy problem, and even that's barely the half of it. Overall, it looks like a possible ultimate theory.

It has, in fact, been dubbed a Theory of Everything by the British theorist John Ellis. That sounds pretentious, but as Ellis tells the story, he was responding to the critics who referred to the theory disparagingly as a theory of nothing. In any case, while the theory has some impressive accomplishments, it is a long road from the things we understand to a complete—or even partial—theory of nature.

A few problems stand out. While the equations of the theory have solutions that resemble the world around us, they have many solutions with drastically different properties: different numbers of dimensions, different numbers of gauge fields, different numbers and types of quarks, leptons, different interactions among the particles. From a theoretical perspective, one of the striking differences is the amount of supersymmetry. The supersymmetry we have proposed to solve the hierarchy problem is the smallest type of supersymmetry one can have in four dimensions. It is interesting because it need not be exact; it can be broken. This is important, since if supersymmetry is a symmetry of nature at all, it can't be exact. There isn't another particle in nature with all the same properties as the electron, for example, except with different spin. With more supersymmetry, breaking is essentially impossible. Broken supersymmetry is complicated, yet rich; unbroken supersymmetry is simple, and boring. I shouldn't exaggerate the boredom; all the phenomena of duality we have described in the previous chapter are understood only in systems that have nonminimal supersymmetry, exploiting all the constraints that the symmetry imposes. But it is still true that the best-understood universes in string theory are the most boring places to live. To put it differently, it is easy for a theorist to understand six or eight dimensions with supersymmetry, but their physics—and chemistry—are not rich enough to produce anything like life as we know it, and probably not stars or galaxies.

But however much we may dislike them, there is no argument that rules out these boring universes. They aren't subject to mathematical inconsistencies (and people have looked) or to some weird dynamics that might cause them to disappear. In fact, the existence of states in string theory that really look similar to what we see around us is highly conjectural. This hasn't stopped me nor my colleagues from writing many papers speculating on a stringy reality. Unfortunately, none of these papers can be said to be making a prediction from string theory. Typically, the author likes one particular solution of string theory or another, and selects one feature that is distinctive and goes beyond the Standard Model. But apart from the arbitrariness of this choice, the author has closed his or her eyes to two huge problems with their proposal.

The most primitive problem is the dark energy, which I will assume is a cosmological constant. Prior to the development of string theory, this was a problem that, however troubling, theorists could ignore. They had a good excuse. In quantum field theory, one can't predict the cosmological constant, as a matter of principle. It's just a number. It may be a very weird number, but there's nothing to be done about it. With string theory, that changes. String theory should be able to predict everything, and it turns out that the cosmological constant is the very first thing it should predict. Already in the early days of string theory, Witten wondered whether something miraculous might happen with the cosmological constant. Perhaps it would just come out zero for some reason.

With unbroken supersymmetry, the string solutions then known had zero cosmological constant. But if supersymmetry is broken, one would expect a cosmological constant set by the energy scale of supersymmetry breaking. Witten set his student Ryan Rohm on this problem. Rohm studied a compactification of one of the superstring theories without supersymmetry. The result was both interesting and disappointing. One *could* compute the cosmological constant. But it was just as large as one naively expected. While Witten has often referred to these miraculous properties of string theory—the cancellation of anomalies, appearance of generations, resolution of puzzles of quantum gravity—there were no miracles here. In the subsequent decades, while many more string theories and many more compactifications have been studied, nothing promising has been found.

Instead, another solution to the cosmological constant problem, of a very different sort, has been put forward. It is one that troubles many physicists, but it has also had a remarkable success. Before introducing any loaded words, let's consider a simple question. Why do humans find themselves on the surface of the earth? The earth is a rather exceptional place. While it's true that we now know there are planets surrounding many stars, and we are starting to develop evidence that many of them might be compatible with the existence of liquid water and other ingredients of life as we know it, the fraction of the universe on the surface of such planets is ridiculously small. Even if there is a solar system like ours around every star, this is perhaps a part in 10^{40}. Most of us would respond: The answer is obvious. Life—even allowing for forms of life very different from those with which we are familiar—almost certainly can't develop in empty space or in or close to stars. It can only happen in these exceptional places like the surface of the earth. Things can't be too hot. One probably needs liquid water. There has to be an adequate supply of heavier elements. And these might be just the bare minimum requirements. So long as there are such planets, though, some fraction will probably host intelligent life, and these are the exceptional places in the universe where life will be found.

Weinberg, following a suggestion by Banks, approached the question of the cosmological constant in a manner related to the question of life on planets. He imagined that the universe is, in some sense, far larger than what we can currently see, and that in different parts of this "metaverse" or "multiverse," the constants of nature, and especially the cosmological constant, take different values. He asked the question: In what parts of this multiverse might one find observers? This is a lot like our question of finding life on earthlike planets, but the question of how often you find planets is a hard one. Instead, Weinberg asked, In universes with laws otherwise just like ours, for what values of the cosmological constant would galaxies and stars exist? That question turns out not to be so hard to answer. The assumption that the laws of nature are the same as ours means that, absent a cosmological constant, it takes a billion years or so after the big bang to form stars. If the cosmological constant is negative, the universe undergoes gravitational collapse, basically becoming something like a huge black hole long before stars and planets form, unless the cosmological constant is extremely small in absolute value, the negative of an *extremely* small number. If the cosmological constant is positive, one has a different problem. Now, unless the cosmological constant is very small, before stars have a chance to form, the universe starts to expand extremely—exponentially—fast, as in inflation. Under these circumstances, the stuff that would normally bind together to form stars never coalesces.

So Weinberg gave up the then popular idea that the cosmological constant (or dark energy) was simply zero and argued that perhaps it should be just small enough that stars could form. Just small enough because this small is already very weird; anything smaller would be even more unlikely. The result is quite close to what was subsequently discovered. Arguably, this was a *prediction* of this amazing discovery. Now, to be honest, in the simplest formulation of this argument, he predicted a cosmological constant about 100 times larger than was subsequently observed, but this is pretty good since:

1. The number we would have predicted by more conventional reasoning is off by 120 orders of magnitude. So one or two is a big improvement.
2. The cosmological constant was subsequently discovered, at a value just a bit smaller than had been searched for previously.
3. The argument is pretty crude. Refinements might account for the difference.

While arguably a big success (not everyone agrees), Weinberg had opened a Pandora's box, both potentially quite interesting but also, in the view of many, threatening to the very process of doing science. This idea that some or all of the laws of nature are as they are because this is necessary for the existence of observers—of life—is known as the anthropic principle. Actually, Weinberg distinguished different forms of this principle. At an extreme, you could take a religious perspective: Some ultimate being set the laws in this way so that humans could exist. But Weinberg had a viewpoint that was, in a sense, very antireligious. Our existence is just an accident. We are not merely an unimaginably small speck in a gigantic universe, but what we think of as the universe is itself just a tiny speck in a universe of universes. We find ourselves in the particular universe we do because it is one of the extremely rare ones that allows for stars and galaxies—and life. He called this the *weak anthropic principle.*

While it is probably true that only a minority of scientists practice a religion, and a still smaller fraction would describe themselves as believers in any fundamentalist sense, even those scientists who describe themselves as personally religious would say they don't allow religion to interfere with their scientific inquiries. They believe they should be studying nature without prejudice. Yet most scientists, certainly most theoretical physicists, believe there is some sort of order in nature, and an underlying simplicity. As Einstein famously said, "One may say the eternal mystery of the world is its comprehensibility." This viewpoint has much historical support in the success of science in accounting for a vast array of phenomena in terms of a small number of principles and compact equations. The *anthropic principle*, if operative, would overthrow this view. Just to explain the cosmological constant would require an absurd number of different universes—10^{120} at least, but far more if other constants of nature—and even the basic structure of nature's laws—are fixed this way. Nature would be absurdly complex, and the very meaning of natural law unclear. For many, it has the feeling that one is giving up on achieving a real understanding. It would be so much more satisfying if we could calculate the cosmological constant, just as we calculate properties of atoms.

String theory, at least in its early phase, had created optimism that it would be somehow possible to understand all the laws of nature, and all the constants of nature, from an underlying structure. Someday, scientists would determine the laws of nature as we observe them, all the constants of nature, and anything else one might want to know. Many had viewed the cosmological constant as some sort of outlier, a problem to be resolved in a much more distant future. But indeed, willingness to contemplate the anthropic principle *does* open a Pandora's box. Perhaps not only the cosmological constant might be determined by anthropic considerations but many or even all the constants of nature, and the laws themselves. The existence of stars depends on many things besides just the universe being sufficiently old. If the strength of the weak force, for example, were very different than it is, stars wouldn't burn, or would burn up too fast. If the electron were much heavier than it is, atoms, molecules, and solid materials would have very different properties than they do, and life as we know it would be impossible. The chaos of possibilities and possible universes gets more and more extreme as one contemplates these different issues.

String theory might possess many possible states. We've talked about different dimensions, different numbers of quarks and leptons, but surely not enough, the orders and orders of magnitude required, to explain the universe in such a fashion. We now have some experience with powers of 10. The sorts of numbers required are 10^{500} at least. We've seen that numbers like trillions, which we throw around in our day-to-day lives, are hard to imagine. This one is inconceivable. If every atom in the universe was itself a little universe, with as many atoms as our universe has, this wouldn't be nearly 10^{500} atoms. Such a mode for explaining the world around us is just too much to swallow. Even with a minimalist view of a creator, this is an extraordinarily complicated way to give rise to life. As a result, my reaction to Weinberg's original proposal went something like: Ha-ha, very clever, but surely nature doesn't work like this.

As skeptical as we were, Tom Banks, Nathan Seiberg, and I considered one possibility that might yield such a vast number. With a combination of disappointment and relief, we established, with some confidence, that this doesn't happen in string theory. We were able to sleep well for a few more years.

Then Joe Polchinski and Raphael Bousso (now at UC Berkeley) came up with a more plausible proposal. We have talked about electric and magnetic charges and have seen that if both are present, each are quantized, by Dirac's monopole argument. This means they take values that can be counted, 1, 2, 3, and so on. Polchinski and Bousso noted that in string theory, there can be many types of magnetic and electric charges occupying the compact dimensions, often hundreds. Each of these can take different values. If each can range from, say, negative five to five, and there are 500 types of charges, there are all together 10^{500} possible configurations of charges. If these numbers are the charges parked in some compact dimensions, then each one represents a different possible universe. So this could be what one needs. Comparing a continuous set of points with a continuum, like an object with a smooth shape, Bousso and Polchinski called this a "discretuum," suggesting something that is almost a continuum but is actually a collection of a vast number of discrete points. Leonard Susskind later called this a landscape.

I was still a skeptic. Tom Banks, his student Lubos Motl (now the host of a conservative Czech blog), and I wrote a paper enumerating a whole set of reasons why Bousso and Polchinski's proposal was likely not realized in string theory. I slept well for a little longer. But then a group at Stanford, looking carefully at certain string models, provided a much more persuasive version of the Bousso-Polchinski story. The group—Shamit Kachru, Renata Kallosh, Andrei Linde, and Sandip Trivedi—were, between them, experts on the relevant string theory, general relativity, and cosmology. They addressed persuasively almost all the objections that Banks, Motl, and I had raised earlier. I, and many others, decided that perhaps this was for real. This work is so famous that it is known by their initials, KKLT. The paper has almost 3,000 citations (Sandip is director of the Tata Institute for Fundamental Research [TIFR] at Mumbai, India).

Accepting that string theory *might* yield a vast landscape of universes, one had to confront the anthropic principle not just for the cosmological constant but for the laws of nature more generally. Banks and I, in collaboration with our postdoc Elie Gorbatov (who went on to a successful career in the financial services industry), considered what this might mean. We realized that, apart from the issue of whether one likes it or not, the anthropic principle faces some real challenges. Weinberg's invocation of the anthropic principle solves the problem of why the cosmological constant is much smaller than otherwise expected. But there are many constants of nature that take small values, which *seem to be of no consequence for the existence of life.* A striking example is provided by the quantity θ of the strong interactions, which we know has to be extremely tiny, less than 10^{-10}. If this quantity were much larger, even about one, this would have no consequence for the universe around us. There are others not quite so dramatic, but similar. So at least for some of the constants of nature, anthropic considerations may not be the answer. Of course, it *could* be that some of these quantities are connected with other quantities of nature that are constrained by anthropic considerations.

Some have despaired. Consider, for example, the hierarchy problem. This might be just like the cosmological constant problem. After all, if we have accepted that there are a vast number of vacua to account for the cosmological constant, among those with small enough cosmological constant, there might be a vast number with different values of the Higgs mass. This Higgs mass might well be selected by anthropic considerations, as we have said. Stars that are hot enough and live long enough to give rise to hospitable planets might require that the Higgs mass be near its present value. In this view, one doesn't expect supersymmetry, or technicolor, or some other explanation for the hierarchy with consequences for the LHC. Discussions of this topic among my colleagues get quite heated.

The whole subject remains a controversial one. Some theorists have argued that the work of KKLT does not convincingly establish that this vast array of states exists (and certainly no one would argue that it *proves* it, in the sense that mathematicians use the word). Tom Banks continues to argue somewhat cogently that these phenomena are not actual properties of string theory.

Some workers, myself included, have at least provisionally taken the viewpoint that the idea of the anthropic landscape *might* be right, and asked where it might lead. We have suspended our skepticism and, in doing so, decided not to adopt an attitude of complete despair. My own approach has been to ask: What are general questions that might have answers typical of some large set of universes in the would-be landscape? One might focus on the possible explanations of dark matter or inflation, for example. I have focused on a simple question, something that must be true of any universe in the landscape that's a candidate for observers. Indeed, the question is much more primitive than Weinberg's. In the constructions of Bousso and Polchinski, most universes as a whole are unstable. Like radioactive particles and nuclei, they will decay. This will typically happen *very* fast, in a tiny fraction of a second. As you might guess, this isn't good. We count on the universe sticking around for a while. Before confronting this very speculative question, let's take on one to which we can give some answers: What will happen to our own universe over the next trillion years or so?

The Fate of Our Universe

In the landscape, universes come and go. This raises the question: What is the fate of our observable universe?

People often ask me whether our knowledge of the universe makes me hopeful about the human condition or leads me to despair. While perhaps too embarrassed to say it outright, the question they are often asking is "Is there a God?" I don't have a satisfactory answer, and I neither want to force people to give up their beliefs nor lead them to some fairy-tale version of the meaning of existence. I have mentioned that Stephen Hawking wore his atheism as a badge of pride. That's fine, but I personally believe, God or not, we need to engage with others in improving the world. Einstein held a view that is close to mine. He certainly didn't believe in an all-powerful being who intervenes in human affairs, but he did marvel at the ability of humans to understand nature. I can't help but feel that there is something about our ability to discern quantum mechanics (or music, art, or literature) that is an element of giving life meaning and value. But Steven Weinberg presents the lessons of science and our knowledge of the universe in a more bleak way, which gives me pause: "The more the universe seems comprehensible, the more it also seems pointless."

As you contemplate the ultimate fate of the universe, I'm

afraid, you may find yourself adopting Weinberg's view, but it is fascinating nonetheless. I first seriously confronted this during a colloquium by an astronomy colleague, Greg Laughlin (for many years at Santa Cruz, now at Yale). His talk, and an article, coauthored with astronomer Fred Adams of the University of Michigan, had the title "A Dying Universe: The Long-Term Fate and Evolution of Astrophysical Objects." The prospects, as the title suggests, are deathly. After our explorations in this book, we look out even further into the future than Greg did. Things don't get better.

In the spirit of powers of 10, we can think about this question on different scales of time. We are now at about 13 billion years since the big bang. As far as humans are concerned, and life more generally, we are in a golden age. In the first few billions of years after the big bang, the universe was not very hospitable to life. But early generations of stars produced large quantities of heavier elements—carbon, oxygen, iron, for example. This debris has become the stuff of new stars but also, we know now, a vast array of planets. Without these elements, life, at least as we know it, could not exist. But in a few billion years, our sun will burn out, and this will be the fate of the stars around us. New stars will continue to form for a while, but star formation will eventually come to an end. When the universe is about 10^{14} years old, the lights will have gone out. What's left are cold stars, mainly white dwarfs. These will occasionally collide, but while the result may include some light, even these encounters will end after a while, about 10^{23} years. In the meantime, the dead stars will gradually peel off from the galaxy.

But worse is yet to come. First, we saw in the chapter on the dark energy that the universe is starting to grow exponentially. At 10^{14} years, the universe has already grown by a huge amount, a factor of 10^{4000} or so, give or take a bit in the exponent. This means the galaxies we know are typically so far from each other as to be invisible. On average, if you were sitting on one atom, the nearest atom would be far too far away to see. More precisely, the remnants of our galaxy are separated from others by unimaginable distances. Space is mostly empty.

It still gets worse. These little islands of dead stars are themselves doomed. We have argued that all matter is radioactive; protons eventually decay. We don't know how long this takes, other than that it's more than about 10^{33} years. Let's suppose it takes about 10^{35} years. The decay of a proton ultimately may produce a positron, a neutrino, and photons. The positrons will annihilate with electrons, producing photons as well. If you could watch this process, it would not be as if it happened all at once. In your neighborhood, the photons are at first very energetic, but as we've seen in the case of the cosmic microwave photons, their energy will decrease as the universe continues to age and they wander off into the nearly empty regions surrounding the galaxies.

So we're left with essentially nothing of the normal matter around us. Actually, there will be a few, *very* low energy photons, a sort of Hawking radiation related to the dark energy. We're also left with the huge black holes at the centers of most galaxies (also known as active galactic nuclei), which will decay by Hawking radiation, converting the bulk of their energy to very low energy. This takes a *very* long time, likely much longer than the time required for proton decay. As for the inventory of energy, more important is the dark matter. Its fate depends on what it is. If it's in the form of WIMPs, many of these will collide with each other over time, converting their energy again into radiation. If the dark matter is in the form of axions, these decay with lifetimes that can easily be far longer than the proton lifetime. The decay products, again, are very low energy photons. At some level, these are details. When the universe is 10^{100} years old, give or take a few factors of a trillion one way or the other, everything is gone; the universe is heat dead.

So the future is depressingly cold and dark. But if there's a landscape, the story is distinctly more complicated. In this case, what we have described so far is the fate of *our* universe. But in a landscape, universes would be born and die constantly. The fact that ours has hung around as long as it has is, in fact, quite remarkable, and almost certainly a clue as to how this all works. In literature and casual conversation, we sometimes speak metaphorically of people as radioactive. The issue is that for most of these universes, the whole universe is radioactive. Most of them decay very quickly after they form. But what does it mean for a universe to be radioactive?

A Radioactive, Unstable Universe

We've seen that there is much in quantum mechanics that is not like our day-to-day experience. One striking phenomenon is known as *tunneling*. Suppose you're hiking and passing over hills as in the figure:

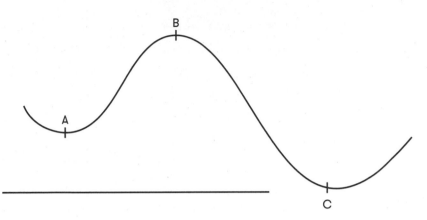

Tunneling

Your destination is point B, but when you reach point A, you stop for a drink and a rest. It's going to be a lot of work to get to the top of the hill above point B. Once you get there, it's an easy trip down.

Now, in quantum mechanics, you don't have to do the work of crossing the hill; you can, literally, tunnel through it. The point is that you can't say with certainty that a quantum mechanical particle, say an electron or an alpha particle, can't be on the top of the hill, even though, if it were a classical particle, it wouldn't have enough energy to get there. The uncertainties in quantum mechanics allow the particle to violate energy conservation for short periods of time. We can think of the particle as burrowing, or tunneling through, the hill. By the time the particle comes out the other side, energy conservation is okay again. This phenomenon is the basis of an array of electronic devices. It is also important in the radioactive decays of radium, studied by the Curies.

Radium has 88 protons and various numbers of neutrons

(various isotopes). For example, one isotope has 138 neutrons and a half-life of about 1,600 years. It decays, emitting a helium nucleus, with two protons and two neutrons, leaving behind a nucleus with two fewer protons and neutrons (radon). The theory of these decays was worked out in 1928 by George Gamow, who played such a crucial role in the development of the big bang theory. He realized that one can think of the large nucleus as an alpha particle bound to the smaller nucleus. To escape, the alpha particle has to climb an energy hill, just as in the figure. For the radium nucleus, the probability of making it through is low, and the decay takes a very long time. Another nucleus that decays this way is plutonium, an important nuclear fuel and by-product of reactions in nuclear reactors. The half-life of plutonium is hundreds of thousands of years. This fact is one of the reasons that disposal of nuclear waste is such a big problem. It is necessary to isolate these materials safely for times that are almost geological.

If the landscape idea is right, our universe has a tiny cosmological constant. As for our particle tunneling through a hill, there are other universes with lower cosmological constant— lower energy—separated by an energy barrier from us. (More properly, we should speak of *states* rather than *universes*. The other universe comes into existence only after the tunneling.) It is like the hiker at point A, having started at point C. It can lower its energy—cosmological constant—if it can get to point B. Classically, it can't. But quantum mechanics is another story. Now, however, we need to know how a *universe* tunnels. How would this happen? The theory of this process was worked out by the late Sidney Coleman, a brilliant, beloved, and somewhat eccentric Harvard professor who, in addition to having a re-markable sense of humor, bore a certain resemblance to Albert Einstein, which he unashamedly exploited. Coleman was not worried about the landscape—he developed these ideas much earlier—but already in quantum field theories like the Standard Model, there was the possibility that our universe might be unstable.

As for the alpha particle, we could get to this other ground state, or "vacuum," if we could just tunnel through the hill. To do this everywhere in space would be essentially impossible, according to Einstein's relativity principle. Faraway regions would have to act in concert simultaneously. But Coleman made an analogy to boiling water. When you heat a pot of wa-ter on the stove, at some point, it is energetically favorable for the water to turn into steam—water vapor. But it doesn't do this throughout the pot all at once. Instead, bubbles of water vapor form, rise to the top, and burst into the air. If you could seal the pot very tightly, the bubbles would collide, gradually turning all the liquid to vapor.

Coleman demonstrated that the same thing happens for a universe as for boiling water. Small regions tunnel through the barrier, and "bubbles"—regions of true vacuum surrounded by false—emerge. Like the bubbles of water vapor, these bubbles grow. In fact, they grow very fast, quickly reaching nearly the speed of light. They collide violently with each other, leaving a universe in the lower energy state, with the debris (particles, possibly hot) from the bubble collisions.

Already in the Standard Model this is a worry. It turns out that, given the mass of the Higgs particle, and assuming that we understand the theory at high energies very well, there is a lower energy/cosmological constant state. Using the procedure Coleman outlined, you can calculate the half-life of the universe. Fortunately, it is many orders of magnitude larger than the present age of the universe. In any case, those who do this calculation have to assume far more knowledge of the properties of the Higgs particle than they have any right to. Still, it's a possibility that we have to acknowledge. As Coleman says in one of his papers on this subject, "Vacuum decay is the ultimate ecological catastrophe; in a new vacuum there are new constants of nature; after vacuum decay, not only is life as we know it impossible, so is chemistry as we know it. However, one could always draw stoic comfort from the possibility that in the course of time the new vacuum would sustain, if not life as we know it, at least some structures capable of knowing joy."

But if the landscape idea is right, this problem—that the universe in which we live is not eternal—is inevitable. Indeed, Weinberg's basic picture is that the fact that our cosmological constant is so small is due to the fact that our universe has been selected from a large number of possible universes with a range of possible values of the cosmological constant, positive and negative. We currently have no idea how, in detail, this would arise. Certainly, if one had a credible theory of the landscape, one might hope to ask questions like: What came before the big bang? After all, our universe might have resulted from the tunneling from a higher-energy universe.

But I've advocated setting aside such challenging questions and just focusing on the half-life of our own universe. As for the worry about the Standard Model, this half-life better be much larger than the present age of the universe. If it was, say, one-tenth as large, the chance that we'd still be here would be nearly zero.

Why should this be? Well, there's a cheap way out. We could try to invoke the anthropic principle again. It would be crucial that we require the existence of people, and perhaps even us specifically. I'd worry that the anthropic principle would then require only that we're alive today, not tomorrow, and so the end, in this case, might well be near. Even more ridiculous, perhaps I would just require that *I* be alive today to observe the universe. Instead, I've advocated an explanation that might actually make the landscape predictive. To understand this, it is helpful to return to Coleman's work. In a paper with a student, Frank De Luccia, Coleman asked about the tunneling problem in general relativity. Here, things were different in two ways. If we start in a universe like ours, with nearly zero cosmological constant, and decay to one with a negative cosmological constant, the universe comes to a catastrophic end, in what general relativists call a singularity (probably something like a black hole). Indeed, I left out part of Coleman's joke before. He writes, after expressing the hope for structures "knowing joy": "This possibility has now been eliminated."

But Coleman and De Luccia discovered another possibility. Under some circumstances, once gravity is included in the mix, the decay does not occur at all. It turns out that if nature is supersymmetric, the condition for a stable universe is satisfied. Nature is certainly not exactly supersymmetric, but it might, in an appropriate sense, be approximately so. With postdoc Guido Festuccia and student Alex Morisse, I showed that with supersymmetry as we conceived it in our discussions of the hierarchy problem, the universe could have an unimaginably long half-life: 10^{100} is a googol; the half-life might be 10^{googol}. This number is known as a googolplex. It is a number I first heard of as a somewhat nerdy junior high school student. Finally, perhaps, it can be put to use.

A Principled Use of the Anthropic Principle?

This would seem to be a sensible use of the anthropic principle, and at the same time a prediction arising from the use of the landscape and the principle. The problem is that this is not, by itself, a prediction of supersymmetry that one might see at the LHC. This requires more input. Several of us have struggled with this question and, at the moment, can't provide an answer one way or the other. It would be exciting if, say, one could predict that supersymmetry might be right around the corner.

The answer might be the hierarchy problem itself. It is possible, as we've indicated, that the strength of the weak interactions is important for life. Since it is the Higgs field—and in essence its mass—which determines this scale, perhaps among the states of the landscape, anthropic considerations select not only low cosmological constant but a small Higgs mass. Perhaps more of the states with low Higgs mass also have a low scale of supersymmetry. While this seems plausible to many theorists, my colleague Scott Thomas, of Rutgers University, and I pointed out that this need not be true. It is possible that other considerations, like the dark matter density, are also selected by anthropic considerations, but the arguments that have been advanced for this are not, at least as yet, convincing.

The other challenges for a landscape viewpoint remain. Ultimately, the situation is quite unsatisfactory. One can simply reject the landscape hypothesis, as many physicists do, as ugly, or unsupported, or even unscientific. But this does not seem, itself, to be scientific. There are questions for which we don't have alternative explanations, and we have theoretical structures (string theory) with at least some of the features required to implement such a hypothesis. Alternatively, one can adopt the landscape viewpoint, but then one has to acknowledge that, at this point in time, we have nothing like a complete theoretical framework in which to make any scientific investigation, and that there are facts hard to reconcile with this viewpoint. I, for one, find this quite unsettling.

CHAPTER 15

ROLLING THE DICE
OF THEORETICAL PHYSICS

We are coming to the end of our journey into some big questions about nature, on scales from the unimaginably large to the impossibly small. In these pages, we have encountered many challenging ideas. I hope readers will forgive me if, at times, I have been too abstruse. I have tried to convey some sense of those features of the universe that are well understood, for which we have good theories well supported by experiment, those for which we have plausible explanations and can hope to study with feasible experiments, and those which are realms of reasonable, or not so reasonable, speculation. I am optimistic that, over the next few decades, humanity will find answers to a number of the questions on our list.

Among these, particle accelerators could solve the mystery of the origin of mass and the hierarchy problem. As the energies physicists probe have grown larger, and the scales of distance smaller, accelerators have become larger and more expensive. Through the latter quarter of the twentieth century, there were only a small number of such facilities worldwide. In the United States, there were particle accelerators at the Brookhaven National Laboratory on Long Island, at Fermilab near Chicago, and at the Stanford Linear Accelerator Center in Menlo Park. Elsewhere in the world, there were accelerators at CERN in Geneva, at the KEK facility not far from Tokyo, and in Beijing.

The United States, which for decades was a leader in accelerator-based high particle physics research, shut off its last large accelerator, located at Fermilab, near Chicago, in 2008. It did not make sense to put large resources into the effort once the LHC was running. Instead, the laboratory became one of several hubs for US activities at CERN. Worldwide, particle physics experiments at this moment are dominated by the LHC in Geneva, Switzerland. This is, in monetary terms, a $10 billion class capital project with an annual budget of about $1 billion (officially, the budget is quoted in Swiss francs). Three large experiments comprise the LHC program. The two larger experiments each involve about 3,000 people. CERN, the European Council for Nuclear Research, is the principal funder of the project and sets its direction. CERN is a consortium with twenty-three member countries and several associates, including Turkey, India, and Pakistan. The United States, as well as the Russian Federation, hold Observer status. The United States pays fees and makes in-kind contributions of equipment and personnel.

The LHC program will continue for at least another two

decades. The accelerator will improve measurements of the Higgs boson, checking carefully whether its properties agree with those predicted by the Standard Model. Searches for supersymmetry and other possible explanations for the hierarchy will continue, with enhanced capabilities. But physicists hunger to go to higher energies. We have seen some arguments that if one could go perhaps another factor of 10 higher in energy, something like supersymmetry might reveal itself. CERN, under the leadership of its current director, Fabiola Gianotti, has plans for a much larger, higher-energy facility, with first stages projected for about 2040. Other large facilities are under consideration in China and Japan. All would be highly international, both in their funding and in their scientific membership. Even sharing these costs, they are very expensive, and in each country, other proposed big science projects compete for funds.

The questions around neutrinos will be the focus of much of the experimental program in the United States in coming years. Neutrinos will be produced at Fermilab and detected in a large detector in the Homestake Mine in South Dakota, now called DUNE, for Deep Underground Neutrino Experiment. We may find support for the possibility that neutrinos played a critical role in producing the asymmetry between matter and antimatter in our universe.

In astrophysics and cosmology, observational and experimental efforts continue to probe the nature of the dark matter and to establish whether the dark energy is truly a cosmological constant. Surveys of galaxies and the cosmic microwave background radiation are providing more and more detailed information about the history of the universe. There are currently some puzzles. There is, for example, a small but persistent discrepancy in various determinations of the age of the universe. This might reflect a systematic problem in some of the experiments, but it may indicate that we are missing some element of cosmic history.

I am optimistic that there are good prospects for progress on the theoretical side. Geographically, theoretical work is more dispersed than accelerator-based physics and large-scale astrophysics and cosmological probes. In the United States, some is at national laboratories, though much is spread over universities throughout the country. This reflects, in part, the fact that theorists are comparatively cheap. Worldwide, there are active efforts in theoretical physics in numerous countries on every continent (except, I say with some caution, Antarctica).

Activities in theory shoot off in a variety of directions. Because the overhead is low, theorists can wander widely. But the chance of any individual theorist having an idea that looks promising, either purely theoretically or as a worthwhile target of an experiment, is not high. Actually having an idea that is true and verified by experiment is a much higher bar, which few theorists will achieve.

One can order activities of theorists in terms of how closely they are connected to planning or interpreting experiments that are happening now, are planned for the future, or are at least plausible for the future. This work can involve providing theoretical support for the development of experiments or for their interpretation. The first class includes calculation of the rates for Standard Model processes at the LHC and possible future accelerators, and the calculation of the chances to detect possibilities like supersymmetry. Many study the prospects for direct and indirect detection of dark matter at present and future experiments. Close to experiment are studies of actual accelerator data, particularly in cases where there appear to be discrepancies with the Standard Model, offering explanations in terms of possible new phenomena in some cases, while in others determining whether or not the Standard Model predictions are truly reliably known. My Santa Cruz colleague, Wolfgang Altmannshofer, is intensely studying some anomalies observed in mesons containing b quarks. Much the same applies to dark matter experiments. My colleague Stefano Profumo, while an enthusiast for a variety of models of dark matter, adopts a view that one must first rule out any possible astrophysical explanation for candidate dark matter discoveries. My colleague Stefania Gori has proposed experiments with the capability to search for exotic forms of dark matter.

Many theorists are engaged in more speculative efforts, and they can, as I've said, roam widely over our list of questions. I expect we will see significant progress on the issue of reconciling gravity and quantum theory, and I believe we are likely to develop a better understanding of the structure of the universe, the significance of the big bang, and the nature of the dark energy. We theorists are people of privilege. I can one day worry about issues of axion production in the early universe; another, the determination of the quark masses; and another, the stability of states in a landscape. I don't have to worry about raising vast sums of money for my research, and a varied research program is likely to improve my job prospects or job security. But my chances on a given day of finding and developing a hypothesis that is interesting and makes sense are not so high. The odds that any such hypothesis will turn out to be true of nature are much lower. But I am grateful for all the luck that has come my way.

NOTES

CHAPTER 1: SURVEYING THE UNIVERSE

14 **Of Newton, he wrote:** Quoted by Abraham Pais, *Subtle Is the Lord: The Science and the Life of Albert Einstein*, 1982, reprint 2005, 14.

CHAPTER 3: WHAT DO WE MEAN BY *UNIVERSE*?

39 **Famously, Stephen Hawking . . . was treated to such a flight in 2007:** You can find this on YouTube, https://www.youtube.com/watch?v=OhIpdSZQZlI.

57 **Astronomers measure the fraction:** There are some discrepancies for particular nuclei, such as lithium. There is debate as to whether this is a failure of the calculations or of measurements of the abundance of these isotopes, or something more significant is going on.

CHAPTER 4: CAN QUANTUM MECHANICS PREDICT THE FUTURE?

65 **The state of the atomic hypothesis:** Quoted by Pais.
65 **Like the Greeks:** Quoted by Pais.
67 **One scientist who worked:** Quoted by Pais.
67 **In 1902, he wrote:** Quoted by Pais.
70 **Of the experience he wrote:** Quoted by Pais.
74 **He has been called:** Graham Farmelo has written a wonderful biography of Dirac with this title. *The Strangest Man: The Hidden Life of Paul Dirac*, 2009, paperback 2011.

74 **Schrodinger's wave function associated:** For those with familiarity with imaginary, or complex numbers, the wave function is actually a complex number; it has a real and an imaginary part. The probability is the sum of the squares of the real part and the imaginary part. This gives rise to many strange effects.

78 **Anderson was dismissive:** Quoted by Pais.

87 **Schrodinger imagined a situation:** I have taken some liberties with Schrodinger's formulation of the problem, but I believe I have captured its essence.

CHAPTER 5: FRUITS OF THE NUCLEAR AGE

98 **Similarly, nature doesn't care:** This assumes you are moving in empty space; if there is a mountain in your way, that may be a problem, but it is not an issue of the underlying laws.

CHAPTER 6: THE WEIGHT OF THE SMALLEST THINGS

122 **This is crucial to nature's alchemy:** I first heard the weak force described as the cosmic alchemist in a television program narrated by science writer Nigel Calder.

122 **This very long time means:** The lifetime is extra long because of the small difference in mass between the neutron and the proton.

CHAPTER 8: WHY IS THERE SOMETHING RATHER THAN NOTHING?

170 **Since that time, the prediction:** This is in part because of better measurements of the Standard Model interaction strengths and in part because it was recognized that unification of the interactions works much better in supersymmetric theories than without supersymmetry.

CHAPTER 9: "THE LARGE NUMBER PROBLEM"

180 **By the uncertainty principle:** In the theory of strong interactions, the ratio of the mass of the proton to the Planck mass is given by a formula of the form mass of proton times an exponentially small number governed by a characteristic number in QCD, which characterizes the strength of the force when two quarks are within a distance of order 10^{-32} cm from each other.

188 **It must be lighter:** There are workarounds where there are no stable particles, but these tend to be in tension with a variety of experimental results and generally don't fit as well with known facts.

191 **This work has had enormous impact:** For his work in these areas, Nathan Seiberg won a MacArthur Fellowship, a Breakthrough Prize, and

other awards. Edward Witten, who already held a MacArthur and a Fields Medal, has more recently won the Breakthrough Prize and the APS Medal for Exceptional Achievement in Research.

CHAPTER 13: CAN WE GET TO A FINAL THEORY WITHOUT GETTING UP FROM OUR CHAIRS?

284 **The next, harder computation:** The correction is typically about 1,000 times smaller than the first term because the various corrections actually involve not only α but are also divided by a factor of 4π.

CHAPTER 14: THE LANDSCAPE OF REALITY

300 **For many, it has the feeling:** David Gross, at a conference, used precisely this phrase, quoting Winston Churchill's exhortations during World War II, "Never, never give in."

305 **But Steven Weinberg presents:** From his book *Dreams of a Final Theory: The Scientist's Search for the Ultimate Laws of Nature*, 1993, Vintage reprint 1994.

ACKNOWLEDGMENTS

Bringing this book from conception to a reality was a process to which I owe a huge debt to many people. A coffee with Robert Irion, who for many years headed the Science Writing Program at the University of California, Santa Cruz, awakened me to what would be necessary to take the germ of an idea and turn it into a book that might appeal to a broad audience. I am enormously grateful to Graham Farmelo, the author of three wonderful books, for his extensive advice on my book proposal, and on writing strategies. My debt is even greater as, once he found the proposal was sufficiently appealing, he introduced me to his agent, Toby Mundy. And to Toby I owe further improvements of the proposal, and with it guidance that led to a much-improved version of the first draft of the book. My editor at Dutton, Stephen Morrow, looked at the proposal and made further insightful suggestions. And he then guided me through extensive revisions. As I said to my children, his many hundreds of comments were payback for all of my criticism of their writing through the years. The result is, I hope, a far more interesting and readable book.

I owe a great debt to many scientific collaborators for all they have taught me about these subjects and who have been critical to those things I have accomplished—among them Ian Affleck, Nima Arkani-Hamed, Tom Banks, Ann Davis, Savas Dimopoulos, Willy Fischler, Dimitra Karabali, Chiara Nappi, the late Ann Nelson, Yossi Nir, Lisa Randall, the late Bunji Sakita, Nathan Seiberg, Yael Shadmi, Yuri Shirman, Mark Srednicki, Leonard Susskind, Scott Thomas, and Ed Witten. The support, of so many kinds, of my Santa Cruz colleagues— George Blumenthal, Sandy Faber, Raja Guhathakurta, Howard Haber, Joel Primack, Hartmut Sadrozinski, Abe Seiden, and so many others—has been crucial to my career as a scientist and teacher, and informs much of this book. Through the years, my postdocs and graduate and undergraduate students provided me with some of the greatest satisfaction and pleasure in my work, and taught me both science and hopefully some skills in communication.

Finally, none of this would have happened without the love and support of my family: my spouse, Melanie Aron, and my children, Aviva, Jeremy, and Shifrah Aron-Dine and Matt Fiedler. Their sharp but gentle criticism of all of my endeavors have helped keep me careful and honest, and kept me focused on what is important in science and in life.

INDEX

INDEX

ABOUT THE AUTHOR

Michael Dine is Distinguished Professor of Physics at the Santa Cruz Institute for Particle Physics, University of California, Santa Cruz. He has been a Sloan Fellow, a Guggenheim Fellow, a Fellow of the American Physical Society, and in 2010 was elected a Fellow of the American Academy of Arts and Sciences. A recipient of the J. J. Sakurai Prize honoring outstanding achievement in particle physics theory, he is also a member of the National Academy of Sciences and served as Chair of the Committee on the Future of Theoretical Particle Physics of the American Physical Society.